Catherine de Courcy is an Irish-born historian, who has worked in Papua New Guinea, as well as Australia and her native Ireland. Her publications include *The Foundation of the National Gallery of Ireland*, *The Zoo Story*, and the popular travel books *Desert Tracks* and *River Tracks*, both co-written with John Johnson. She also contributed to the international survey, *Zoo and Aquarium History: Ancient Animal Collections to Zoological Gardens.*

Catherine was the Melbourne Zoo's official historian from 1988 to 2003. This book is an edited version of her 1991 M.A. thesis, and includes a cd-rom containing substantial additional material – maps, illustrations, and a digest of the Zoo's minute books from 1861 to 1964.

Royalties from the sale of this book will be donated to the Horticultural Department at the Melbourne Zoo.

This edition is limited to 200 copies.

Copy no. 181

Evolution of a Zoo

A History of the Melbourne Zoological Gardens

1857-1900

Catherine de Courcy

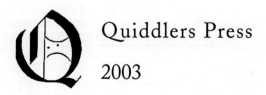 Quiddlers Press

2003

Quiddlers Press
P.O. Box 3034
Auburn, Vic. 3123
www.quiddlers.com.au

© Catherine de Courcy, 2003

National Library of Australia
Cataloguing in Publication data:

De Courcy, Catherine, 1957– .
Evolution of a zoo : a history of the Melbourne Zoological Gardens, 1857-1900.

Bibliography.
Includes index.
ISBN 0-9581949-1-2

1. Royal Melbourne Zoological Gardens – History – 19th century. 2. Zoos – Victoria – Melbourne – History – 19th century. I. Title.

590.739451

The illustration on the front cover is an undated photograph, probably early twentieth century, by W.H. Dudley Le Soeuf, in the Le Soeuf Family Papers.

Designed and typeset by Quiddlers Press
Printed by Aristoc Offset, Glen Waverley, Vic.

Contents

Chapter 1	Foundation of a Zoo	1
Chapter 2	The Utilitarian Zoo	16
Chapter 3	The Australian Zoo	37
Chapter 4	The International Zoo	53
Conclusion	The Legacy of the Nineteenth Century	76
Appendices		79
Bibliography		83

Author's Dedication

This book is an edited version of my University of Melbourne M.A. thesis, which was written in 1989-1990 in honour of the late Dr Alfred Dunbavin Butcher (1915-1990), member of the Board of the Royal Zoological Society of Victoria from 1948 to 1987, and chairman of the Board from 1962 to 1987. As a gift to Dr Butcher on his retirement, the Board established a fund to encourage a graduate student to write a history of the Zoo. He was tremendously enthusiastic about the project. Regrettably, Dr Butcher died three months before my thesis was finished. It was completed, and is now published, in his memory.

September 2003

Acknowledgements

I wish to acknowledge the valuable assistance given to me by many people in preparing this work. Primarily, I must thank my thesis supervisor, Dr John Foster, for his constructive criticism and timely advice. The generosity of Nick le Souef, Mr and Mrs David Wilkie, and Mrs Purvis, all descendants of nineteenth century Melbourne Zoo staff, in allowing me access to family archival material was appreciated. Professor Rod Home and Dr Linden Gillbank of the Department of History and Philosophy of Science, University of Melbourne, both provided me with useful leads to resources, for which I am grateful. The administrative staff of the Melbourne Zoo, most notably, Graham Shotter, former Director of the Zoo, Di Cardwell and Lesley Vink, were all willing to assist me in every way possible.

The personal and moral support which I received form my family, friends and colleagues made it possible for me to pursue this project; I particularly wish to thank J.W. de Courcy, Marlene le Brun, Elizabeth Wallace, Ian Rodgers and, most especially, John Johnson.

August 1991

Abbreviations

ADB	*Australian Dictionary of Biography*
ASV	Acclimatisation Society of Victoria
VPP	Victorian Parliamentary Papers
ZASV	Zoological and Acclimatisation Society of Victoria
ZSL	Zoological Society of London
ZSV	Zoological Society of Victoria

1. Foundation of a Zoo

The official date of the foundation of the Melbourne Zoo is 6 October 1857. Yet it was not until 1870 that any characteristics of a zoo in the modern tradition were introduced. The activities of the Zoo's original organisers in the intervening period were not those usually associated with modern zoos. This raises some fundamental questions as to what the motives and objectives of the founders of the Melbourne Zoo were, what they achieved and why the creation of the Zoo was delayed for so many years.

6 October 1857 was the day of the inaugural meeting of the Zoological Society of Victoria (ZSV). The conveners had intended to establish an ornithological society. As soon as the meeting opened, however, they broadened the scope of their interest and founded a zoological society 'both for the purposes of science and for that of affording the public the advantage of studying the habits of the animal creation in properly arranged zoological gardens.'[1] Some weeks later they published a full list of objectives.[2] Various interests were represented in the list, including the importation of song-birds, but the objective which commanded their immediate attention was the creation of a zoo.[3]

The concept of a zoological garden for public entertainment spread following the foundation and success of the Zoological Society of London (ZSL) gardens in Regent's Park in 1826. Widely regarded as the third zoo to be established in the modern tradition,[4] the Regent's Park Zoo became one of the more fashionable places in London in the 1830s. An elephant, an Indian rhinoceros, a giraffe and other exotic animals were purchased to attract visitors.[5] Access was restricted to members and friends only, yet the crowds were such that it was of-

---

1. ZSV minutes, 6 October 1857.
2. Appendix I, Objectives of the Zoological Society of Victoria.
3. The abbreviation 'zoo' was first used around 1847 in Britain; it entered popular usage in the late 1860s in a music hall song, 'Walking in the Zoo on Sunday' by The Great Vance: 'The Stilton, Sir, the cheese, the O.K. thing to do / On a Sunday afternoon is to toddle in the Zoo' (quoted by Blunt, *The ark in the park*, p.29). In Melbourne the word became widely used in the 1880s.
4. *Schönbrunn* in Vienna (1752), and the *Jardin Des Plantes* (1789) are considered to be the first of the modern zoos; both were founded on the basis of royal menageries.
5. Blunt, *The ark in the park*, pp.41, 80, 163; Desmond, 'The making of institutional zoology', p.230.

ten difficult to get into the gardens, particularly on Sundays.[1] The London Zoo was used as a model in cities throughout the world.

Civic pride was the motivation behind the establishment of many of these zoos which followed in the wake of the ZSL's success. Prominent citizens, believing that their city ought to have a zoological garden, were the chief promoters of the new institutions;[2] this attitude was evident at the foundation meeting of the Melbourne Zoo. The city of Melbourne had undergone significant social and economic changes during the gold-rush of the 1850s. The small pastoral town had become a major colonial city, its population rising from 77,345 in 1851 to 410,766 in 1857. Cultural features of provincial cities in Britain were established in the 1850s, amongst them the University of Melbourne (1853), the museum of natural history (1853), and the Public Library which incorporated an art collection (1856). Learned societies were organised to provide forums for discussion of a wide range of issues; the Philosophical Institute, modelled on the Royal Society in Britain, was the most important of these: founded in 1855 by a merger of the Philosophical Society of Victoria and the Victorian Institute for the Advancement of Science, both of which had been founded in 1854, it became the Royal Society of Victoria in 1860.

The foundation of the ZSV was undoubtedly a result of the same civic enthusiasm. The profile of the original committee was one of professional men, most of whom had no apparent scientific knowledge of zoology. It included a general medical practitioner, an actuary and several lawyers as well as merchants; many of them also had pastoral interests in the colony.[3] Foremost amongst them was Thomas Black, the medical practitioner, who first suggested that a zoological society was preferable to an ornithological society, believing that the former 'had long been a desideratum in the Colony.' Black had moved from Sydney to Victoria in 1843 and, after establishing himself in private practice, had become associated with community activities both inside and outside his profession. He was treasurer of the first medical association (1846-1851) and was also credited with inspiring the establishment of the Bank of Victoria.[4] He was to remain an active and influential member of the ZSV for many years.

---

1. Blunt, *The ark in the park*, pp.32ff.
2. Mullan & Marvin, *Zoo culture*, p.110.
3. Appendix II, Foundation members of the Zoological Society of Victoria.
4. 'Garryowen', *The chronicles of early Melbourne*, pp. 325 and 888.

Foundation of a Zoo

The other objectives which featured on the ZSV's published list reflected the particular interests of individuals among the founding committee. One founder with well defined interests who was to have a significant influence on the course of the ZSV was the owner and erstwhile editor of the *Argus*, Edward Wilson (1813-1878). After a series of unsuccessful ventures in the 1840s, Wilson took up journalism. He was a forceful writer, often attacking La Trobe's Government and criticising the squatters' monopoly on the pastoral land in the colony. In the months prior to the foundation of the ZSV Wilson expressed an interest in developing the agricultural resources of Victoria. He advocated the creation of a State agricultural body which would establish experimental farms and gardens, pointing out that no Government body had yet shown any interest in experimenting with the introduction of animals which might adapt to the climate and become additional resources for the colony. He organised experiments which involved moving fish from the Murray River to the more accessible Yarra river, and began arrangements for the importation of British song-birds to Victoria. The objective relating to the importation of song-birds in the ZSV's list can probably be attributed to him directly.[1]

While civic and utilitarian interests featured prominently in the ZSV's published objectives, it is worth noting the lack of importance accorded to the role of science in the new organisation. None of the usual features of a British specialised scientific society – for example research, public discussions, lectures or published proceedings – was included in the ZSV's list of objectives. These omissions are curious, particularly when considered with reference to the development of learned societies in Britain and the complex origins of the Zoological Society of London. Although the gardens in Regent's Park might have appeared to have been the ZSL's principal activity, they were in fact a subsidiary part of that institution.

The ZSL was fundamentally a scientific society, one of several specialised societies which emerged in the late eighteenth and in the early decades of the nineteenth centuries. It was founded in 1826 and, after a brief power struggle, the scientific interests of career zoologists gained dominance over the interests of the gentry. The ZSL organised forums to discuss questions and undertake experiments in animal physiology, it corresponded with members around the world, imported rare and useful animals, and prepared zoological reports. Regu-

---

1. *ADB*, v. VI, pp. 412ff.; Wilson, 'On the Murray River cod', and 'On the introduction of the British Song Bird'.

lar meetings were held and the papers read and published in the *Proceedings*. The Committee's meetings attracted the best young zoologists in the city, and were often attended by people who were about to sail on survey expeditions. (Darwin attended a meeting, two months prior to his departure in the Beagle in 1831.) The popularity of the Gardens made the ZSL one of the wealthiest societies in London but, to the zoologists, they were merely the means by which they could pursue their scientific interests.[1]

None of the founders of the ZSV could be regarded as career zoologists. W.A. Archer and F.M. Selwyn, the two founding members who had an expressed interest in zoology, did not promote a scientific approach to the subject. Archer, a statistician who arrived in Melbourne in 1852, had a keen interest in botany and zoology and was elected a Fellow of the Linnean Society. Selwyn, who lived in Brighton and was a partner in a legal practice in Little Collins Street, possessed what was considered to be 'the best collection of animals to be found in Victoria' and for this reason was selected as the first president of the ZSV.[2]

If the Zoological Society of Victoria had adopted the same course as that of the Zoological Society of London, it would have had access to enough raw material to support significant contributions to the study of zoology. There were several important areas of scientific research concerning Australian fauna which had not yet been resolved. There was an enormous interest in Australian zoology in Britain, particularly with regard to marsupials and monotremes, both of which confounded the classification schemes already in operation.[3] The problems associated with collecting and preparing animal specimens and amassing data about animals in their natural habitat for transmission to Britain made the study of the zoology of Australia potentially very rewarding. The creation of the ZSV could have provided an ideal platform from which the colonial scientist might have contributed to the research but, by excluding lectures, discussions,

---

1. Desmond, pp. 227-228, 234-235; Mitchell, *Centenary history*, pp.112-113. Much of the interest of the gentry lay in the farm attached to the ZSL which was to breed game for restocking their depleted estates; it was attacked as a symbol of amateurism by the career zoologists and disposed of in 1823.
2. *ADB*, v. III, pp.41-43. ZSV minutes, 6 October 1857, 8 January 1858. Selwyn stepped down from this position in deference to Redmond Barry in January 1858; he died in 1859.
3. Kathleen G. Dugan, 'The zoological exploration of the Australian Region and its impact on biological theory', in Reingold & Rothenberg, p.80.

publications or similar activities, no provision was made to promote these studies within the organisation.

Some of the omissions may be understood. The Museum in the University of Melbourne was being expanded to include zoological specimens. In addition the ZSV members may have considered the Philosophical Institute an adequate forum for zoological debate. It often addressed zoological issues at its meetings and published the papers in its *Proceedings*. Although several founding members of the ZSV were also members of the Institute, few of the authors responsible for these papers had any involvement with the ZSV. Edward Wilson was the one exception; in 1857 he read papers on his Murray cod experiment and on the value of introducing British song-birds to the colony. He even organised a sub-committee of the Institute to consider the importation of song-birds.[1]

However, there were more fundamental reasons for the lack of scientific interest in the ZSV. Hoare, in his study of Victorian societies in the 1850s, identified some of the difficulties which the founders of learned societies had to face: 'Amateurism; a shortage of skilled men; the layman's obstinacy and the utilitarian emphases of a pioneer society'.[2] The ZSV, unlike the Zoological Society of London, did not have a group of men with a knowledge of zoology to call upon although there were several scientists in the colony who might have been valuable assets.[3] William Blandowski and Frederick Mc Coy, the Government zoologist and the foundation Professor of Natural History in the University of Melbourne respectively, were both scientists whose official positions suggested that they would have had interest in the formation of a zoological society. But each was engrossed in other activities at the time. Blandowski was going through a stormy period in his professional career which culminated in his permanent departure from the colony early in 1859; before he left, however, he attended one meeting of the ZSV, joined one of the deputations to Government and promised to donate some rare animals.[4] Frederick McCoy was spending much of his time developing the collections of the Museum in the University despite formidable opposition from the Philosophical Institute of Victoria.[5]

---

1. The sub-committee survived for several years but, unlike other sub- committees, it never produced any reports.
2. Hoare, 'Learned societies', p.26.
3. Many members of the ZSL had been active in the zoological club of the Linnean Society from 1822-1826 (Desmond, p.157).
4. *ADB*, v. III, pp.182-183; ZSV minutes, 2 November 1857; *Argus*, 6 November 1857.
5. Kohlstedt, 'Australian museums of natural history', p.6.

Foundation of a Zoo

Even the potential source of unusual animal specimens which would inevitably emerge from a zoo was not enough to distract him from his other work at this stage.

The ZSV's pursuit of a recreational facility based on a well known model created by one of the foremost scientific societies in London may have been another reason why Blandowski and McCoy did not express interest in the new organisation. Learned societies and scientists in the colonies were under pressure to cater to utilitarian needs of the society. As Hoare points out, with reference to the Philosophical Institute of Victoria, colonial societies had 'to tread a precarious road between the 'useful' and the 'abstract' sides of science.'[1] The demands on Australian science and societies to develop local resources were powerful and the support for a purely intellectual approach weak. As a result colonial science in the mid-nineteenth century was localised and utilitarian.[2] Undoubtedly, the ZSV founders planned to be of service to some sections of the community and intended to allow their gardens to be used as a recovery station for recently imported stock.[3] But, initially, the creation of a zoo was their only interest.

The plan was to establish the rudiments of a zoological garden with the aid of Government as quickly as possible in order to attract subscribers. Their fees were to comprise the major source of funds to maintain the daily operation of the Zoo. The inducement to the public to pay an annual two guinea subscription was a card which entitled each member, accompanied by two ladies, free access to the Zoo and to all of its exhibitions. The general public was to be given restricted access to the grounds at designated times on payment of an entrance fee.[4]

The Government was co-operative at first. A 30 acre site on the north side of the Yarra in the vicinity of the Botanic Gardens was granted by the Chief Secretary, William Haines.[5] He also promised a sum of £3,000 and, on the basis of that pledge, work on fencing the paddock in preparation for the acquisition of animals commenced early in 1858; most of this work was complete by April. The wildlife collection began with a large donation of animals and birds from

---

1. Hoare, 'Learned societies', p.11.
2. Ian Inkster & Jan Todd, 'Support for scientific enterprise, 1850-1900' in Home (ed.), *Australian science in the making*, p.103.
3. Appendix I, Objectives of the Zoological Society of Victoria.
4. ZSV minutes, 29 May 1858; *Argus*, 1 March 1858.
5. Pescott, *The Royal Botanic Gardens, Melbourne*; *Argus*, 27 October 1857; notes on map presented to the Chief Secretary by a deputation of the ZSV (PROV, 1189, 744, A7755).

the ZSV's president, F. M. Selwyn. Except for two monkeys, the donation was composed entirely of Australian fauna and included kangaroos, wallabies, a laughing jackass, emus, eagles, black swans and a native companion. They were temporarily placed in the Botanic Gardens under the care of the Government Botanist and Director of the Gardens, Ferdinand von Mueller. Other donations were pledged: Blandowski promised some rare, unspecified animals which he had collected in the colony, the Philosophical Institute's birds and aviary were also to be given to the ZSV and a Mr James Cox of Tasmania promised a dozen deer, which arrived in October.[1]

In addition to these donations, plans were made to acquire some foreign animals to please the crowds. In May 1858 the ZSV opened negotiations with a Mr Billings, proprietor of Wombwell's menagerie.[2] The arrangement was that Billings would move his menagerie to the new zoological gardens, and would take up a position as managing keeper. The ZSV was to have the option of purchasing the animals and, in return, it would pay for their keep and provide accommodation for Billings. The negotiations appear to have been successful and the Zoo was scheduled to open to the public in August 1858.[3] At this point, George Coppin joined the ZSV. Coppin was involved in several successful entertainment venues in Melbourne including two theatres, the Cremorne Amusement Park and four hotels.[4] His experience in commercial enterprises promised to be of great value in promoting the entertainment value of the proposed Zoo. But he never got the opportunity to demonstrate his ability because some weeks later financial problems altered the course of the ZSV.

The original financial arrangement with the Government was vague. Haines had pledged the sum of £3,000 but he was removed from office before the money was paid.[5] After some negotiation the new Chief Secretary, John O Shanassy, agreed to honour Haines' promise but insisted that the vote should be

---

1. *Argus*, 6 November 1857; ZSV minutes, 8 April 1858, 29 May 1858; Zoological Committee minutes, 3 July 1858, 8 October 1858.
2. The menagerie probably included lions and monkeys, both of which were easy to acquire in Melbourne at the time; for example, lions were offered for sale from a ship to the ZSV at its second meeting (*Argus*, 3 November 1857). There is no reference to Billings of Wombwell's menagerie in Mark St Leon's *The Circus in Australia*.
3. ZSV minutes, 11 May 1858.
4. *ADB*, v. III, pp.459ff.
5. Serle, *The golden age*, p.261.

Foundation of a Zoo

approved by the House of Parliament.[1] By this stage the ZSV had spent over £1,000 of borrowed money on fencing Richmond Paddock and on printing circulars.[2] Its collection of animals was still being looked after in the Botanic Gardens although Mueller was becoming increasingly concerned about the poor conditions in which the animals were confined.[3]

A rapid solution was necessary. Several ideas to solve the problem were considered and, after consulting Henry Barkly, the Governor and patron of the ZSV, a proposal was arrived at which essentially gave the Government a greater role in the establishment and operation of the new zoological gardens. In the course of a meeting with the Chief Secretary, the ZSV's representatives offered to hand the project over to the Government entirely but suggested that would be no further delays in opening the Zoo if they were permitted to have some part in its management there.[4] They also claimed that the Zoo could be self-financing even though, by this stage, they had only received £39 in subscriptions, another £100 in pledges, and a list of sixty potential subscribers. No attempt seems to have been made to emphasise utilitarian activities as the representatives made it clear that the creation and opening of the Gardens were of paramount importance, and the potential financial independence of the Gardens was the only benefit offered at this stage. In response, the Government established a Board of Management for the zoo comprising five members of the ZSV and five Government appointees, some of whom were drawn from the ranks of the Board of Science at the ZSV's request.[5]

The composition of the Zoo's Board of Management had a significant impact on the future direction of the institution. Some of the Government appointees were obvious choices. Capt Charles Pasley, R.E., for example, as the Commissioner of Public Works would be involved with fencing and construction work in the Gardens. Ferdinand von Mueller was already involved with the

---

1. Note on letter, 26 April 1858 (PROV, 1189,745).
2. ZSV minutes, 29 May 1858.
3. Correspondence from Capt Stoney to the Chief Secretary, 24 April 1858 (PROV, 1189,745).
4. ZSV minutes, 29 May 1858.
5. Appendix III, Members of the first Board of Management of the zoological collection. The Board of Science had been set up early in 1858 to advise the Government on 'all matters wherein special scientific or technical knowledge in requisite' (Gazette, 3 June 1858); its members included Frederick McCoy, Ferdinand von Mueller, ARC Selwyn, the Director of the Geological Survey, Thomas Skilling, the director of the model farm, as well as political and public service nominees. The Board was dissolved in 1860; much of its activity had been concerned with mining.

Foundation of a Zoo

ZSV in that he was caring for its animals in the Botanic Gardens. Frederick McCoy's position as foundation professor of natural history in the University of Melbourne and his association with the Natural History Museum made him another logical choice for the Board. But the decision to appoint Thomas Embling was a most interesting one.

Embling was a member of the Legislative Assembly, a medical practitioner and a man described by Serle as 'busy in every good cause.'[1] His diverse interests included reforms in the Yarra Bend Lunatic Asylum, support for the eight hour day movement and advocacy of penny banks, bathing, northern exploration and a transcontinental railway.[2] Of more immediate significance was his interest in the introduction of unusual agricultural animals to Victoria on a large scale. In 1856 he organised and chaired two parliamentary select committees to enquire into the acclimatisation of animals. The first of these committees was concerned with bringing alpacas to Victoria. It investigated the different types of alpaca and their qualities, and discussed methods of smuggling the animals out of their native habitats in South America.[3] Embling's second committee investigated the importation of livestock generally, arguing that 'advantage would be taken of the present lack of animals in Victoria, and that active measures would be resorted to to introduce not only the best of the British breeds, but also varieties of new stock from other lands.' Animals which were considered suitable for importation included the angora goat, the quagga, the pheasant and partridge, various deer and, again, the alpaca. An annual grant of £3,000 was requested for this purpose.[4] However, the reports were not well received in Parliament and Embling withdrew them.[5] The invitation to Embling from the Government to represent it on the Zoo's management board now gave him another opportunity to pursue this interest.

As the Government-appointed Board of Management took responsibility for the new Zoo, the ZSV was effectively disbanded. On the suggestion of McCoy, the Zoo and the Botanic Gardens were united. He considered the site

---

1. Serle, *The golden age*, p.255
2. *ADB*, v.IV, pp.140-141.
3. Smuggling was necessary because most of the countries in the region prohibited the export of the animals. (Report from the Select Committee of the Legislative Council on the Alpaca', VPP 1855-6, no. D11a, pp.1, 4).
4. 'Report from the Select Committee of the Legislative Assembly upon Live Stock Importation', VPP 1856-57.
5. Gillbank, 'The origins of the Acclimatisation Society of Victoria', p.366.

Foundation of a Zoo

allocated to the zoological collection unhealthy and recommended that the animals should be accommodated on the other side of the river in the Botanic Gardens.[1] The amalgamation facilitated this arrangement as well as making the operation of the Zoo economical and convenient.

Mueller, as secretary to the Board, wrote to the Chief Secretary requesting ratification of the plan. He explained that both animals and visitors would be sheltered from the sun and the wind if the animals were placed around the Botanic Gardens; he also pointed out that 'the appearance of an European Zoological Garden may at once be obtained without having to wait for years for the pleasure grounds to grow into shape' He further suggested that the land already fenced in might be handed over to the Botanic Garden and the ZSV suitably reimbursed. The request was approved[2] and Mueller was given charge of organising the zoological collection under the direction of the Board of Management, which now called itself the 'Zoological Committee.'

With the establishment of the zoological gardens the ZSV had achieved its primary aim but, by becoming a part of the Government funded Botanic Gardens, it had lost its independence. Money was returned to subscribers and all of its papers and documents were transferred to the Botanic Gardens. The former ZSV members now on the Zoological Committee were fully supportive of the developments. The implications of being answerable to Government emerged during the first meetings of the Zoological Committee. Embling and, D.S. Campbell, another member of Parliament sitting on the Board, both made it clear that Government funding was available for the purchase and maintenance of useful animals only. These animals were to be held in the Botanic Gardens where they might form an exhibit of some interest; however, the primary purpose of holding them there was to allow them to breed in sufficient numbers before being distributed around the colony. The arrangement with Billings was abandoned and a four horned sheep, lent by him to the ZSV, was returned. When George Coppin offered lions for sale, claiming that he was fulfilling an outstanding order, the Zoological Committee refused to purchase them. Over the following three years alpacas, sheep, deer, kangaroos, emus, koalas, various game and song birds and other animals were held in the Zoo in the Botanic Gardens while zebras, an orang outan and other exotic animals associated with public menageries were rejected. Before long, the zoological collection had more

---
1. Zoological Committee minutes, 24 July 1858.
2. 9th, 13 October 1858, (PROV, 1189,745,G8503).

in common with a farm of slightly unusual animals than with a nineteenth century European zoo.[1]

Mueller's attitude to the zoological collection reflected the character of the colonial scientists' brief in the mid-nineteenth century, described by Inkster and Todd as 'a product of colonial needs as perceived by and pressed upon governments.'[2] In his reports of the progress of work in the joint gardens, he stressed the usefulness of the animals which were being cared for at public expense. For example, in his monthly report for June 1860 he commented that the high number of births among the llamas showed 'that whilst we prove these creatures to be well adapted to this country, we can also show a valuable increase of public property on this establishment.'[3]

Mueller, with Thomas Embling, was virtually running every aspect of the Zoo. Few members of the Zoological Committee attended the monthly meetings; often meetings were cancelled and occasionally they were held without a quorum. Mueller and Embling attended frequently and, consequently, dominated the direction and operation of the Zoo during this period. Embling's influence was particularly noticeable in discussions about animal acquisitions. At his suggestion many of the animals mentioned in the 1856 Legislative Assembly reports on animal importation were now considered for acquisition; furthermore, he did much of the work in locating sources and negotiating deals. He entered into communication with the colonial Government in South Africa on the subject of acquiring domesticated South African animals, including the ostrich, the eland and the quagga.[4] Other negotiations in which he took a direct part, and probably initiated, included an application to the American Consul for beavers, and a request for hares and partridges from India.[5]

Mueller introduced the Zoological Committee to a more lucrative source of animals, the international exchange network through which surplus Australian animals were sent on exchange to zoological gardens abroad.[6] In 1859 and 1860 the Zoological Committee opened the exchange transactions by sending pairs of

---

1. Zoological Committee minutes, 6 December 1858, 1 December 1859, 8 December 1859.
2. Inkster & Todd, in Home (ed.), *Australian science in the making*, p.113.
3. PROV, 1189, 748.
4. His list of desirable South African animals was sent to the Chief Secretary of the Cape Colony with a sum of money (Zoological Committee minutes, 8 December 1859).
5. 'Annual report of the Government Botanist and Director of the Botanic and Zoologic Garden, 1860-1861', VPP 1860-1, no. 19, p.8; Zoological Committee minutes, 3 January 1860.
6. Zoological Committee minutes, 14 December 1858.

Foundation of a Zoo

black swans to various gardens around the world including the Botanic Garden of Copenhagen, the Zoological Garden of Cologne, the Botanic Garden of Java, the Calcutta Botanic Garden and the Acclimatisation Society in Paris. The London Zoological Society, which was sent numerous pairs of black swans during this period, was also given several pairs of eagles.[1] Some time later the Zoological Committee began to receive animals in return; most of them were small game animals including hares, rabbits, fowl doves and water birds.[2] The lines of communication were now open and were to remain so for many years.

The local network was also a useful source of animals. People living in Victoria and in nearby colonies donated Australian and other animals to the zoological collection. Porcupines, owls, canaries, wallabies, opossums, turtles and koalas were amongst the animals received from Australian sources.[3] There was seldom an exchange or monetary reward involved with these donations and many of the individuals had no further direct contact with the Zoological Committee. Their generosity was a reflection of the widespread interest in natural history in the mid-Victorian period. In the late 1850s the Botanic Gardens and other cultural institutions were popular places to visit.[4] David Fleming, in a study of colonial science, argues that 'the practical associations of natural history were greatly enhanced by its appearing to be the intellectual aspect of pioneering'; as a result, natural history reconnaissance became 'an acceptable style of scientific endeavour in the new societies themselves'.[5] Mueller had already established a network of collectors around the colony. Some were young botanists being trained under his direction, others were amateur collectors whose work was appreciated and rewarded by Mueller.[6] He may well have used his established local network to the advantage of the zoological collection; on at least one occasion he offered seeds and plants to a donor in exchange for a kangaroo.[7]

---

1. 'Annual report of the Government Botanist', p.10.
2. A consignment of animals from Cologne included hares, rabbits, fowl, doves and water birds in exchange for two black swans – Report from Dr Mueller to the Chief Secretary, July 1860 (PROV 1189, 748).
3. 'Annual report of the Government Botanist', pp.10-11.
4. In 1858-59, an estimated 200,000 visitors to the Botanic Gardens; Museum, University of Melbourne recorded 32,000 the same year (Inkster & Todd, in Home (ed.), *Australian science in the making*, p.107).
5. David Fleming quoted in Home (ed.), *Australian science in the making*, p.ix.
6. A.M. Lucas, 'Baron von Mueller: protégé turned patron', in Home (ed.), *Australian science in the making*, p.141.
7. Zoological Committee minutes, 5 November 1860.

Foundation of a Zoo

The Zoological Committee also acquired animals from ships' captains who had brought them from South Africa, India and Ceylon; some were received as donations, others were purchased. Individuals living abroad also sent gifts to the Victorian Gardens. Edward Wilson, a founding member of the ZSV, was in England by the time the Government set up the Board of Management. Nevertheless, his interest in the project was enormous and he maintained constant contact with Mueller and the zoological collection. He promoted the general interests of the Zoological Gardens and managed to solicit donations of animals from wealthy landowners in Britain.[1] In addition, he organised several consignments of game birds for transportation to Australia, some of which were purchased with funds voted by Victorian Legislative Assembly.[2]

However, the largest consignments of birds sent to Victoria by Wilson consisted of song-birds, reflecting his continuing personal interest in the creatures. Thrushes, blackbirds, starlings, sparrows and other birds were transported in wire cages and without an attendant.[3] Many died on the way and those that survived usually arrived in very poor condition with little plumage and covered in insects. Mueller treated them with olive oil and placed in the palm house as protection from the cold. On recovery the birds were transferred to the large aviary where they were encouraged to nest and breed.[4]

The Zoological Committee's activities were limited to organising local and international transfer of animals and maintaining the collection in the Botanic Gardens. The influence of experienced scientists on the Committee did not extend to encouraging an analytical approach to the work. Consequently, the use of the zoological collection remained within the framework of the colonial scientific tradition. Indigenous animals were sent abroad for research and study, usually to the Zoological Society of London. Dingoes, water hens, fish, ducks and magpies were amongst the animals which followed the earlier batches of black swans and eagles to the ZSL in the early 1860s. There is no evidence in the Melbourne Zoo archive to suggest that the ZSL maintained a scientific correspondence with any member of the Zoological Committee, nor did it attempt

---

1. For example, Lord Caernarvon arranged with Edward Wilson to send some animals from his estate (Zoological Committee minutes, 15 January 1859).
2. Zoological Committee minutes, 9 March 1859.
3. 'Annual report of the Government Botanist', p.9.
4. Zoological Committee minutes, 4 May 1860.

to solicit scientific data from this source.[1] Communication between the two institutions appears to have been concerned mostly with arranging animal exchanges. The results were not always satisfactory for one or other party; occasionally the ZSL did not get what it asked for; in February 1861 it sent a request for 'menura, leipoa, Phascolarctos, Petragole and Casarca, and any waterfowl but black swans.' However, Edward Wilson, who had just returned from Britain, suggested that the black swans should be sent because they were 'still prized at home for effecting easy breeding.'[2]

The steady pace of the Zoological Committee under the direction of Mueller and Embling was interrupted in September 1859 when an incident occurred in the Botanic Gardens. Magpies donated by Embling to the collection escaped from the aviary and, according to Mueller, became 'very destructive' to small birds in the Gardens. Much to Embling's annoyance, Mueller had them shot.[3] From then on animosity between the two men developed and the atmosphere at Committee meetings was tense. Rumours about the condition of the health of some of the animals began to circulate; in response Mueller asked the entire Zoological Committee to inspect the collection. No fault in his care or treatment of the animals could be found.[4] In July 1860 another attempt was made to undermine his position when Embling suggested that a permanent chairman should be appointed to the Zoological Committee. Again, he countered this action successfully and the motion was postponed. However, he was aware that, while some members of the Committee supported him, 'leading men in the community' wanted the collection transferred to Royal Park where it would be outside his control.[5] For a while he fought against this move arguing that the cost involved in moving the animals would include additional salaries and new

---

1. The Australian zoology community was small; George Bennett, a Sydney physician, contributed to the proceedings of the ZSL; he earned a reputation in British zoological circles for observations, scientific papers and specimens. Much of his information was channelled through Richard Owen (Dugan, 'The zoological exploration of the Australian Region', in Reingold & Rothenberg, pp.90-91). There was also a Victorian zoologist of note, P.H. MacGillivray, a medical practitioner and an active invertebrate zoologist (Lucas, 'Baron von Mueller', in Home (ed.), *Australian science in the making*,p.144); he produced illustrated descriptions of Australian polyzoa, 'sea-mosses' (Inkster & Todd, in Home (ed.), *Australian science in the making*, p.113). MacGillivray never became involved with the ZSV.
2. Zoological Committee minutes, 18 February 1861.
3. Zoological Committee minutes, 2 September 1859.
4. Report from Dr Mueller to the Chief Secretary, August 1860 (PROV, 1189, 748).
5. Report from Dr Mueller to the Chief Secretary, July 1860.

Foundation of a Zoo

fences; finally he suggested that no decision should be made until Edward Wilson returned from Europe as he would be able to give advice based on the latest developments in zoological gardens abroad.[1] His request was agreed to and, as anticipated, Wilson returned at the end of the year full of ideas influenced by European trends. Within weeks of his arrival in Melbourne he began to put plans into action which were to alter the course of the Zoo yet again.

Wilson's return from Europe marks the end of the first phase in the history of the Melbourne Zoo. The founders of the Zoo achieved their fundamental objective when the Government agreed to establish a zoological collection in the Botanic Gardens. Although the plan to create an entertainment centre with exotic and unusual animals was not achieved at this stage, the rudiments of the future Zoo had been established. The utilitarian principles of the Government appointees to the Zoological Committee gained a level of dominance that was almost inevitable for the period; a zoological collection for recreational and scientific purposes was probably premature in a colony that was still trying to establish its primary industries. As a result, all of the animals in the original Melbourne Zoo collection were chosen because they were considered suitable for distribution around the farms of the colony or for the purposes of exchange and not because the public might find them curious or educational. This attitude was to remain dominant for the next decade.

---

1. Report from Dr Mueller to the Chief Secretary, July 1860. The following month Mueller confirmed that he wanted the zoological collection removed from the Botanic Garden. His decision was made with reluctance; the additional workload had not troubled him but, given the problems of working with the Zoological Committee, he decided to return to his 'original and higher position of being in every way directly responsible only to the honourable the Chief Secretary and left free to act under the sole control of his office.' (Report from Dr Mueller to the Chief Secretary, August 1860.)

2. The Utilitarian Zoo

 The best known aspect in the history of the Melbourne Zoo is its acclimatisation era. Indeed the founding of the Acclimatisation Society of Victoria (ASV) in 1861 is sometimes seen as the real beginning of the history of the zoo. Clearly this is misleading in view of the preceding discussion. This chapter retraces ground already covered by historians, however it is important to identify the association between the Zoo and the ASV in order to understand how each related to the other.

In December 1860, shortly after he returned to the colony, Edward Wilson joined the Zoological Committee which was, by then, concentrating almost entirely on the mechanics of importing and caring for useful animals on a large scale. Herds of llamas, angora goats and deer had been acquired and were breeding successfully in the Botanic Gardens. The song-birds were even more prolific and already some had been liberated at various locations around the colony. Game birds were also liberated but in controlled environments where their progress could be monitored carefully. Meanwhile negotiations for a consignment of alpacas from South America were underway and many other animals had been promised from all over the world.[1]

While the operations of the Zoological Committee were functioning smoothly, the tension amongst its members was barely suppressed. As Mueller anticipated, however, Edward Wilson devised a solution whereby the work could continue without being over-reliant on the Government Botanist. It also gave him the opportunity to introduce some of the ideas regarding animal introduction which had impressed him on his recent trip to Europe.

The focus of Wilson's plan was the establishment of the Acclimatisation Society of Victoria (ASV) devoted to the importation and naturalisation of animals. The inaugural meeting of the ASV was held in February 1861 and presided over by Sir Henry Barkly. He opened the proceedings with a speech discussing the value of introducing foreign animals in Australia. He referred to individuals who had imported animals in the past and suggested that a united effort was much more likely to be successful in the long run. Wilson then took the floor and picked up on this theme. He began by stating that he knew of 'plenty

---

1. 'Annual report of the Government Botanist', pp. 7-11.

of instances in which gentlemen had tried experiments, had failed, and, having lost money in this manner, had thenceforward devoted their means to something else.' He envisaged the creation of an organisation which could assist private acclimatisers in their endeavours by providing a centre for information and discussion on experiments: 'Let them not go stumbling along in desultory experiments, making mistakes and yet not gathering any information from those mistakes' (*Argus*, 26 February 1861).

The plan fell into place when Wilson outlined the administrative details of the new organisation. Financially, it was to be supported by private subscription and, initially, the managing authority was to be composed of the existing Zoological committee with a few additional members. Tactfully he announced at the inaugural meeting that he did not wish to supersede the Committee, but 'was anxious to strengthen their hands, and to give a fresh expansion to their functions,' adding that he was 'exceedingly anxious that there should be nothing said or done that might in the slightest degree hurt [the Committee members'] feelings' (*Argus*, 26 February 1861). By drawing all of the Committee into the management of the ASV he avoided upsetting any of them and also ensured that the acclimatisation work already underway would continue.

Although the idea of the private subscription was to make the ASV independent, Wilson fully realised that the new organisation would need Government aid in the early stages if it was to succeed. He received a generous response when he called upon it for assistance and, until 1864, the ASV received several substantial grants from the public purse as well as a forty acre site in Royal Park.[1]

The Zoological Committee continued to work until the Council of the ASV was ready to assume full control.[2] As well as organising further exchanges and purchases, it began the process of transferring animals from the Botanic Gardens to their new accommodation in Royal Park. Camels belonging to the Committee had been grazing in the Park for some time prior to the advent of the ASV. They had been allowed to roam but, as the rest of the collection began to arrive, their freedom was curtailed. As if by way of compensation, Mueller

---

1. Grants received for 1861-1864 totalled £9410 ('Acclimatization Society[sic], Return to an order to 31st July 1864', VPP (1864-5), no. C12); subscriptions for the same period were almost £2600 (*Second annual report of the ASV*, 1863, pp.12-13, and *Third Annual Report of the Acclimatisation Society of Victoria*, 1864, p.30). Royal Park was placed into the care of Trustees for the use of the ASV (*Government Gazette*, 25 March 1862, p.529); the zoological collection in the Botanic Gardens was also vested with the Trustees (ASV minutes, 9 April 1862).
2. The last recorded meeting of the Zoological Committee was on the 14 Aug 1861.

provided saddles for them so that, at least, they could be given some exercise. Towards the end of the year a police van was remodelled to move the animals through the city. They were sent in batches as the work on each enclosure was completed; the Chinese and Aden sheep, llamas and angoras were amongst the first to be transferred. By 1863 most of the animals had been moved although, two years later, Mueller was still trying to get rid of a few which had been left behind.[1]

The ASV planned to do more than simply provide an exhibition of its animals in Royal Park. The focus of its objectives, which covered a range of activities, was the introduction, acclimatisation and domestication of useful or ornamental innoxious animals.[2] Sir Henry Barkly, a keen supporter of the concept, presented some practical ideas as to how to pursue this objective: 'collect the experience of past failures...take care that shipments are made in sufficient quantities to give a fair chance to the experiment and that proper care and attendance for the animals are provided on board...investigate the nature of the climate and soil and the natural conditions with respect to wood and water...and also see which of the residents in [selected] districts are most likely to give attention and assistance.' He added that the ASV should undertake these duties for private individuals as well as for itself.[3]

Although ornamental animals were mentioned, the tenor of the objectives was undoubtedly utilitarian. The objectives referred neither to the exhibition of animals nor to the 'rare and uncommon species' which had appeared in the ZSV's list in 1857. The colonial zoo, in this case, had become a farm of unusual agricultural animals held on the presumption that they would be of use to Victoria's pastoralists. Popular, urban interests were not considered and, if members of the public wanted to look at these animals, they were welcome to visit but they were not going to be wooed with the promise of curiosity and entertainment.

Under the guidance of Wilson, the ASV had moved away from the London Zoo model to one which combined features of the French and the British acclimatisation societies. Most of the ASV's objectives had been copied directly from the British Acclimatisation Society's list which, in turn, had been based on the

---

1. ASV minutes, 12 February 1861, 1 May 1861, 20 November 1861, 12 September 1865; *The Yeoman and Australian Aclimatiser*, 2 November 1861.
2. Appendix IV, Objectives of the Acclimatisation Society of Victoria.
3. *First annual report of the Acclimatisation Society of Victoria*, 1862, p.28.

French acclimatisation society's objectives.[1] The *Société zoologique d'acclimatation*, founded in Paris in 1854, was the first organisation to be completely devoted to the introduction and breeding of foreign animals.[2] Its first president, Isidore Geoffroy St Hilaire, was regarded in Melbourne as being 'the father of modern acclimatisation.'[3] St Hilaire summarised the objectives of the Société in his inaugural address as 'nothing less than to people our fields, our forests, and our rivers with new guests; to increase and vary our alimentary resources, and to create other economical or additional products.'[4]

Although the British and the Victorian objectives were almost identical, there was a fundamental difference between the perceived acclimatisation needs of the two localities. This difference was recognized several years before the ASV was founded and was described succinctly by Embling's Legislative Assembly Committees in 1856. Britain, one Committee observed, was 'too full' for the beneficial introduction of new species, while Victoria was 'a virgin land...whose soil, climate and capabilities are unsurpassed, if equalled, which has no animal fitted for man's use, but the few kinds imported, and which has many millions of broad acres lying utterly waste and unproductive.'[5] More specifically; 'some 28,000,000 acres, or nearly one-half [of Victoria], are... yielding nothing whatever to the general revenue of the Colony.'[6] The perception of Victoria as vast and under-utilised was at the basis of the proposals for both animal and plant introduction in the colony during this period.[7]

The difference in the needs of Britain and Victoria was reflected in the approach which the respective acclimatisation societies took to accomplish their objectives. The British Acclimatisation Society worked on the premise that if the additions of 'really valuable animals and plants, useful for food and manufacture...are to be important, they must, as a logical consequence, be few.' It considered that the most likely way to achieve its goals was to undertake some carefully

---

1. Buckland, *The acclimatisation of animals*, p.31. Only two of the ASV's objectives, those referring to international information exchange and to the award system, were original.
2. Osborne, 'The collaborative dimension of the European Empires', p.5.
3. Comment by T.J. Sumner during the first annual meeting, *Argus*, 26 November 1862.
4. Paraphrased by Buckland, p.7. Much of the French acclimatisation interest was based in its colonies in North Africa (Osborne, p.5) where the *Société* encouraged the introduction of the alpaca, the kangaroo, various game, fish and silkworm (Buckland, p.8).
5. 'Report ... upon Live Stock Importation', p.4; 'Report ... on the Alpaca', p.iv.
6. 'Report ... upon Live Stock Importation', p.3.
7. Gillbank, pp.364-365, discusses the attempts to improve agriculture in the colony prior to the foundation of the ASV.

chosen experiments, pursue them thoroughly and then extend them on the basis of the results. Its Victorian counterpart took a radically different approach, largely on the basis of Edward Wilson's advice. Wilson believed that speedy work was necessary in the colonies: 'I do not much believe in the expediency of slowness of action...Life is short, and if we are to be useful in our generation we must keep moving.' Basic needs of a colonial society did not force the pace; Wilson claimed: 'So far from our wants being greater than yours [ie. Britain's], they seem to be rather less, and considerably more amply provided for...We are, in fact, about the best-fed people in the world.' He explained that altruism was the motivation: 'we are anxious to do justice to vast opportunities...a boundless area is opening before us day by day, and we feel ashamed of letting it lie idle while swarms of animal life are panting to get access to it.'[1]

Discussion and promotion of the theory of acclimatisation was an important part of the work of the ASV in the early years. Several papers on the subject were published by it and distributed widely. In all of these the European influence was evident. The first such statement was a paper originally read by Frank Buckland to the Royal Society of Arts in November 1860. Buckland, a British naturalist well known for his articles in the *Field, Land and Water*, had initiated the foundation of the British Acclimatisation Society. He was known for his interest in zoophagy and his rapport with the staff of several zoos, including the London Zoo, helped to maintain his supply of exotic meats; elephant trunk, rhinoceros and giraffe were amongst the London Zoo stock for which he devised recipes. In 1859 he attended a dinner organised by Richard Owen to introduce the eland as a potential source of food in England and an attractive addition to the parks. He was particularly impressed by a speech given at the dinner by David Mitchell, the secretary of the Zoological Society of London, during which Mitchell suggested the formation of a society to acclimatise the eland and other useful animals. The following year Buckland founded the 'Society for the Acclimatisation of Animals, Birds, Fishes, Insects and Vegetables within the United Kingdom.'[2]

Buckland's interest lay in increasing the variety of food available in Britain; in compiling his list of desiderata, he used the London Zoo collection and comments by Mitchell to guide him. The eland, antelopes and deer featured

---

1. Correspondence in the *Field* and the *Times*, reprinted in the *Yeoman and Australian Acclimatiser*, 13 September 1862.
2. Buckland, pp.5-7, 31; Barber, *The heyday of natural history*, pp.144-146.

The Utilitarian Zoo

prominently on his list while the alpaca, pheasants and other game birds were also discussed. Another animal on the list was the kangaroo, which Buckland recommended because it was a hardy animal which bred well, needed little attention, particularly in the parks of the midland or southern counties, and was an excellent meat. But the concern which took up much of his time was acclimatisation of fish and it was this interest which was to afford the greatest contact between the British Acclimatisation Society and the ASV during the 1860s. Buckland's paper on acclimatisation was a comprehensive statement of the objectives of the British Acclimatisation Society and was a useful medium through which to introduce the Victorians to the European acclimatisation movement.[1]

The first important statement emanating directly from the ASV was written by Frederick McCoy and read at the first annual general meeting in 1862;[2] it was also the closest the ASV came to presenting a philosophy of acclimatisation. In his address McCoy presented a rationale which was fundamentally creationist. He subscribed to the law of representative forms, whereby areas with similar climates but separated by natural obstacles were inhabited not by the same animals but by 'representative species'. According to the theory, these species had been originally placed in the middle of the district they inhabit and probably as a single pair. Their 'specific distinctiveness' meant that they could not have had a direct connection with similar species in other parts of the world.[3] By way of example McCoy laid several specimens from the collection in the University Museum before the audience and discussed their similarities; the specimens included a barn-owl and a kestrel-hawk from Europe, and an owl and hawk native to Victoria. After citing several other examples of representative species, although no further examples of Victorian animals, he concluded by examining the distribution of ruminants 'good for food' around the world: Africa and India, he noted, each had about fifty suitable ruminants, North and South America had about twelve different species in this category, but Australia did not have a single one.[4]

The implications of his argument for the acclimatiser were clearly outlined: 'the acclimatiser may...bring with the absolute certainty of success all the repre-

---

1. Buckland, pp.11, 19, 26-27.
2. Frederick McCoy, 'Anniversary address ... on acclimatisation, its nature and applicability to Victoria' in *First annual report of the Acclimatisation Society of Victoria*, pp.33-51.
3. McCoy, 'Anniversary address', p.37.
4. McCoy, 'Anniversary address', p.39.

sentative species of any group into each of the localities.'[1] The work of the acclimatiser, he declared, was 'this great piece of nature's work ...left to us to do' and he expressed his pleasure that Australia had been colonised by Englishmen because, 'if Australia had been colonised by any of the lazy nations of the earth, this nakedness of the land would have been indeed an oppressive misfortune.'[2]

McCoy's attitude to acclimatisation ignored the discussions surrounding the theory of natural selection proposed by Charles Darwin in his *Origin of the Species* (first published in 1859). Darwin had addressed the issue of acclimatisation briefly in the *Origin* by considering 'how much of the acclimatisation of species to any peculiar climate is due to mere habit, and how much to the natural selection of varieties having different innate constitutions, and how much to both means combined.' He suggested that while habit played a considerable part in modification of the constitution and structure of a species, 'the effect has often been largely combined with, and sometimes overmastered by, the natural selection of innate variations.'[3]

Even though the section in the *Origin* on acclimatisation was clearly defined, the lack of interest in Darwin's opinion indicates that the latest scientific theories were not discussed with reference to the work which the ASV planned to undertake. Even after Darwin's theories became the subject of a widespread debate in the colony, the ASV did not alter its approach to its work.[4] An awareness on the part of the ASV of the possibility of changes developing in introduced animals as they adapted to their new habitat, even over a protracted period of time, would surely have encouraged the Council at least to monitor the development of its stock closely. However, both McCoy and Mueller, the most influential and experienced scientists on the Council, opposed the views of Darwin and spoke out publicly against his theories.[5] Wilson, on the other hand, is described by Barry Butcher as 'an admirer of Darwin' and, on his return to England, became a close neighbour of Darwin. But, even in his communications from Britain, he did not encourage a more sophisticated approach to acclimatisation.

---

1. McCoy, 'Anniversary address', p.37.
2. McCoy, 'Anniversary address', pp.36-37.
3. Darwin, *Origin of the Species*, pp.137-139.
4. The first, full-scale public debate was initiated in 1863 by George Halford, first medical professor at the University of Melbourne (Barry Butcher, in Home (ed.), *Australian science in the making*, p.154).
5. Butcher, p.167, and Moyal (ed.), *Scientists in nineteenth century Australia*, p.187.

The Council did not remain unaware of some of the implications of the theory of natural selection during its decade of acclimatisation activity. Dr Henry Madden, a member of the ASV Council, cited Darwin when he challenged a French theory on cross breeding and the Angora goat. The ASV had received some advice from the French acclimatisation society which recommended breeding the angora goat with the common goat every 4th or 5th generation in order to prevent 'the degeneracy necessarily caused by in-breeding.' Madden claimed that after the introduction of 'inferior blood' it was impossible to return to a pure strain. To support his comments, he quoted Darwin's theory that 'all organic beings are capable of modification in form, under the influence of selection' in order to illustrate that cross-breeding was a new experiment and should be treated as such. Madden did not condemn cross-breeding but suggested that the breeder should be aware of the potential results of both cross-breeding and in-breeding and should monitor the results very carefully.[1] Despite its interest in Madden's opinion, the Council did not appear to deviate from its basic approach of transferring an organism from one country to another, moving it around the colony, and measuring success by whether it bred.

At a practical level, the ASV Council was continuing the work of the defunct Zoological Committee. The international connections first established by the Zoological Committee were developed by the ASV as it actively sought to broaden its contacts abroad through various forms of advertising. Circulars were distributed among sea captains and shipping companies, lists of desirable animals with the price the ASV was willing to pay for them were posted in major shipping ports around the world, notices were inserted in newspapers in major foreign cities encouraging gifts and purchases, and all correspondence included a request for 'animals and birds of an economically useful character, and those native birds already domesticated in your locality whether for ornament, song or use.' In addition, the British Government encouraged its emissaries to gather information about acclimatisation in the countries of their postings, and to support the work of the acclimatisers in the colonies including Victoria. At the annual meeting in November 1864, Sir C H Darling, stated that he was including references to the ASV in his communiques to London, with information about animals and vegetables most suitable for acclimatisation in Britain and Europe.

---

1. *On the true principles of breeding developed in letters by Chas H, MacKnight and Dr Henry Madden*, pp.7, 31, 35, 41.

The Utilitarian Zoo

And he also stated that he had asked for reports to be sent to the ASV from British diplomatic representatives abroad.[1]

Botanic gardens, zoological gardens, acclimatisation societies and friendly individuals entered into communication with the ASV and for a number of years there was no shortage of suppliers of animals, particularly game.[2] In return the Council sent Australian fauna, mostly acquired by donation from Victorian settlers, to its contacts around the world. Black swans, kangaroos and emus were particularly popular, as were various pigeons and other birds; they were usually sent off in pairs, although sometimes kangaroos were sent one at a time; these were probably destined for display in a zoological garden. The Council ruled that only Australian fauna should be used for exchange purposes and, despite requests from abroad for animals imported to Victoria, it adhered to this policy rigidly.[3]

Mueller's role in maintaining the contact was still very important. On one occasion, for example, he was asked if he would approach his contacts to acquire birds after it was found that the ASV's agent in Calcutta, Gillanders and Arbuthnot, could not fill an order. On another occasion he informed the Council that he had sent black swans to Japan in the hope of inducing an exchange. Sometimes, however, he received letters from abroad requesting an exchange of animals and, much to the annoyance of the Council, he merely informed them of the interest and held the correspondence in his own files. In one such instance, after Mueller withheld communications from Moscow and Palermo, the Council insisted that all correspondence concerning acclimatisation should be directed to the ASV office on receipt.[4]

The animals which the ASV received both on request or by gift fell into three broad categories: firstly, animals whose primary function was perceived as economic; secondly, those which were seen as a combination of game and food;

---

1. ASV minutes, 12 June 1861, 17 June 1862; 'Circular issued by the Council', appendix to the *Third Annual Report of the Acclimatisation Society of Victoria*, pp.33-34; 'Proceedings at the Third Annual Meeting', *Third Annual Report of the Acclimatisation Society of Victoria*, p.26.
2. In 1865, for example, the ASV had sent animals to London, Paris, St Petersburg, Amsterdam, Rotterdam, Hamburgh, Cologne, Copenhagen, Calcutta, Mauritius, Bourbon, Sicily, Rangoon, and Java as well as the other Australian colonies and New Zealand. (*Fourth annual report of the Acclimatisation Society of Victoria*, 1865, pp.66-67).
3. ASV minutes, 22 July 1862; the policy statement was made in response to plan by Mueller to send imported animals abroad on exchange.
4. ASV minutes, 17 May 1864, 24 January 1865, 5 May 1868.

and thirdly, those which were brought in to enliven the natural habitat of the colony as well as having some useful auxiliary function. The animals in each category were chosen for similar reasons, and their acquisition, breeding, distribution and, often demise, followed a pattern. These activities dominated the work of the Council until 1870.

Animals in the economic category included goats, camels, sheep, silkworm, ostriches and bees. Wool-bearing goats, reputed to survive in harsh conditions similar to those in Victoria, occupied much of the Council's agenda. Of these, the herd of Angora goats may be cited as an example of the pattern followed by the ASV in this category. The Council looked upon the Angora goat as 'a valuable addition to the permanent wealth of the colony.'[1] Using evidence based on a small number of goats imported earlier by the Zoological Committee and living in Royal Park, the Council observed their heavy fleeces of wool and considered them admirably suited to the climate. In addition the Angoras could apparently subsist on 'scanty herbage' and, therefore, were able to 'bring into practical utility land which for sheep is absolutely useless.'[2]

In 1866 ninety-three Angora goats arrived in Melbourne. They had been ordered from Constantinople through a Mr Philpott, the ASV's agent in London, a process which had taken over a year and cost over £1,000.[3] The animals were transported in the charge of a man called Thatcher who made the voyage between England and Melbourne several times in the 1860s for both the ASV and for the recipients of Australian animals in London. On this particular trip he also brought ostriches and other birds as a gift from the *Société zoologique d'acclimatation*. The arrangement with Thatcher was not unusual as there was often a high mortality rate amongst animals travelling unaccompanied on the the long sea journey between Europe and Australia.[4]

As far as the Council was concerned, the Angora goats fared moderately well in Royal Park and, in 1869, in an effort to relieve the ASV's poor financial situation, attempts were made to sell them. But, to the surprise of the Council, there was little public interest in the animals and none was sold. Eventually Samuel Wilson, a wealthy pastoralist with an extensive property in the Wim-

---

1. *Third annual report of the Acclimatisation Society of Victoria*, p.31.
2. *Fourth annual report of the Acclimatisation Society of Victoria*, p.8.
3. ASV minutes, 16 November 1864, 23 January 1866, and *Fifth annual report of the Acclimatisation Society of Victoria*, 1867, p.5.
4. *First annual report of the Acclimatisation Society of Victoria*, p.28.

mera and a keen supporter of the ASV, agreed to care for them. His comments on their condition when they arrived in the Western District perhaps explain some of the reasons why there was so little interest in them: 'they were in a most delicate state, and had not, apparently, before been on pastures which suited them.' In return for caring for them, Wilson was entitled to a portion of the increase and produce. After three years with Wilson the goats had improved in both quality and quantity and some were sold. In 1876 another batch was sold to provide funds for the zoological collection. Eventually, in 1881, as Samuel Wilson was making arrangements to retire to Britain, the ASV had to resume custody of the entire herd. Again, there was little public interest in the animals. Further reasons why farmers were not enthusiastic about them emerged; these included their tendency to roam, the need to shear them twice a year, their sensitivity to the cold and the problems of separating them from common goats. Undoubtedly many farmers were aware of the difficulties and decided against taking responsibility for the goats. But, during the period in which the goats had been on Wilson's property the gardens in Royal Park had been transformed into a zoological garden and there were no facilities for holding them. Eventually the ASV was forced to sell the animals at a low price. However, a few of the finest specimens were kept in the collection.[1]

The ASV's experience with other animals introduced to promote economic growth followed a similar pattern. Different types of wool producing goats, camels, silkworm, Ligurian bees and ostriches were supported with enthusiasm, money and energy in the beginning but, for a variety of reasons, each project was eventually abandoned.[2] If reproduction of several generations was the measure of success, as it was in the *Société zoologique d'acclimatation*,[3] some of these animals

---

1. ASV minutes, 30 March 1874, 13 April 1874, 6 December 1876, 1 February 1881, 2 May 1881; C.M. Officer 'Angora goats', corrigenda to F.R. Godfrey, 'The history and progress of the Zoological and Acclimatisation Society of Victoria', in *Annual Report of the Zoogical and Acclimatisation Society of Victoria*, 1900, p. 22.
2. Much to Mueller's annoyance, the camels were sold in 1865; he wanted them to be used on the *Ladies Leichardt expedition* (ASV minutes, 10 October 1865); the Cashmere goats and the alpacas were deemed unsuccessful in 1867 (*Fifth annual report of the Acclimatisation Society of Victoria*, pp.10-11); the Ligurian bee mixed with the common honey bee and rendered that experiment unsuccessful by 1868 (Thomas Black, letter, *Argus*, 23 May 1868); sericulture was reluctantly given up in 1872 although the ASV promised to give assistance to anyone who wished to continue with the experiments ('Report of the Council' in *Proceedings of the Zoological and Acclimatisation Society of Victoria*, Vol. I, 1872, p. 9].
3. Buckland, pp.8-9, discussed the award system of the French Society.

were moderately more successful in acclimatising than others, but none fulfilled the economic ambitions of the founding members of the ASV.

The Council's relationship with its game animals followed a different pattern. These animals, which included various types of deer, partridge, quail and hares, were introduced in smaller numbers with the noble intention of providing 'for manly sports, which will lead the Australian youth to seek their recreation on the river's bank and mountain side rather than in the cafe and casino.'[1] Shooting and hunting were popular pastimes in Victoria; many bored sportsmen went so far as to import deer, partridge and other game because Australian animals were not considered sufficiently interesting to sustain the demands of the colonial gentleman hunter.[2]

Thomas Austin was one of the more successful of the private acclimatisers; he bred rabbits, hares, pheasants and other game on his property at Barwon Park. His shooting parties were renowned in the colony; the Duke of Edinburgh joined two such parties in 1867 and 1869 respectively, during which he shot hundreds of animals and birds.[3] Austin was also generous with his rabbits, distributing them in breeding pairs to friends around the colony.[4] On one occasion he sent a dozen rabbits to Woods Point, east of Melbourne, in response to a request from the ASV which had, in turn, received a request from the town's borough council.[5] A life member of the ASV, Austin corresponded with the Council, sometimes to share advice, sometimes to organise an exchange of stock. The ASV did not express any interest in the introduction of rabbits; it never imported any on its own account, although it did care for some silver grey rabbits left in Victoria by Edward Wilson.[6]

The game animals were the most popular of the ASV's introductions amongst its members. Unfortunately they were also popular amongst illegal hunters and introduced predators such as feral cats. These obstacles threatened to hamper the attempts to encourage game breeding in the bush. For a number of years, the game animals were kept in Royal Park where breeding habits were carefully monitored. In 1864, however, it was decided that the theory of accli-

---

1. 'Circular issued by the Council', p.34.
2. Rolls, *They all ran wild*, p.23.
3. Rolls, *They all ran wild*, pp.34-35.
4. Rolls, *They all ran wild*, pp.28-29.
5. ASV minutes, 22 August 1865.
6. ASV minutes, 14 July 1863.

The Utilitarian Zoo

matisation demanded that animals should be released from the confinement of the Park to havens in the bush as soon as they were healthy enough to survive. The birds and hares were sent around the colony, usually in breeding pairs, as gifts, on exchange or sold by tender to subscribers, all on condition that they would be kept in the colony. Deer and other quadrupeds were occasionally distributed in small numbers to ASV subscribers and to acclimatisation societies in places such as Ballarat and Tower Hill. Various animals were also delivered by ship to remote spots in Victoria and to isolated islands in the hope that they would breed and provide food for future victims of shipwreck; in 1865, for example, rabbits, geese, goats, pigs and fowl were released on the Auckland Islands.[1]

However, the Council realised at an early stage that, in order to acclimatise game in the sort of quantities which might assure its survival despite the predators, safe rural reserves were necessary. Phillip Island and, later, Gembrook to the east of Melbourne, were chosen as sanctuaries. The idea was to remove the animals as far away from habitation as possible. F.R. Godfrey, in explaining the need to establish the reserve at Gembrook, said that 'the Council had hitherto made a mistake by liberating birds near Melbourne, where they fell a prey to boys, hawks and unsportsmanlike persons, who shot anything and everything they could;' if the game were to be placed in an isolated area,' there was no doubt that they would thrive, so that in a few years there would be enough for sport and for food.'[2]

Animals were first sent to Phillip Island in 1862. The advantage of the Island was that it was free of feral cats and had a very small population. John Mc Haffy, one of the few people living there, became the ASV's honorary supervisor. For the next seven years large numbers of animals, including hares, pheasants, partridges, quail, and ducks, were sent to the relative security of the Island to breed. Both Mc Haffy and the keeper appointed to care for the animals sent regular reports about developments back to Melbourne. Members of the Coun-

---

1. ASV minutes, 6 September 1864, 21 September 1869.*Fifth annual report of the Acclimatisation Society of Victoria*, p.30; a visitor to the Islands in 1868 reported that the animals were thriving (*Age*, 29 May 1868). Other isolated breeding grounds included Wilson's Promontory, Sandstone and Churchill Islands (*Fourth annual report of the Acclimatisation Society of Victoria*, pp.65-66).
2. Godfrey, a pastoralist and business man, and vice-chairman of the Aboriginals Protection Board, was a long-serving member of the ASV and the ZASV, and president of the latter on several occasions (*ADB*, v.IV,, 1972; *Proceedings of the Zoological and Acclimatisation Society of Victoria*, Vol. I, p.11).

cil also visited the Island on occasion. For a number of years the system worked well and the animals bred successfully. In September 1865, for example, Mc Haffy reported that the Californian quail and hares were multiplying 'with extraordinary rapidity.' Unfortunately, Phillip Island was not entirely safe from hunters who visited the island ostensibly to shoot native game, a legal activity in certain seasons, and there was little the Council could do to stop them.[1]

However, the big threat to the future of the ASV's sanctuary emerged in 1866 when the Government decided to sell part of the Island. By November 1868 a large section of it had been settled and Mc Haffy reported increasing difficulty in protecting the animals and birds. Some months later the Council decided to end its connection with Phillip Island after it received a report which stated that the settlers on the island had 'clubbed together for the entire destruction of the deer and hares belonging to the Society and now liberated on the Island.' When the majority of the settlers later claimed to support the work of the ASV, the final decision to quit the Island had already been made. The animals were rounded up and sold to friends of the ASV at low price.[2]

A number of years later the ASV established another reserve on Government land at Gembrook. The zoological collection was expanding and the ASV's new secretary, Albert le Souef, was finding that breeding and rearing pheasants in Royal Park was 'most anxious, troublesome and expensive.' On his suggestion, the ASV purchased some land in the vicinity of Gembrook which was described as an extensive forest containing 'innumerable fern tree gullies abounding in insect life' and suitable for breeding all game birds. Le Souef consulted Edward Wilson before he put his proposal to the Council and had received a pledge from him of £100 a year to support the project.[3]

Acclimatisation at Gembrook had a promising beginning. For a number of years le Souef, who visited the area regularly, reported satisfactory progress in the birds' adaptation to the new environment. However various attempts to eradicate the rabbit, by now a pest causing havoc throughout the colony, spread to Gembrook. Foxes and cats turned loose to eat rabbits also attacked the birds of Gembrook. Then poison, which was regarded as a cheap and effective

---

1. ASV minutes, 25 February 1862, 3 February 1863, 12 September 1865, 7 January 1868. The Council was informed that it had no authority to prevent the hunting of native animals on Phillip Island, ASV minutes, 11 February 1868.
2. ASV minutes, 3 November 1868, 29 June 1869, 13 July 1869.
3. ASV minutes, 27 January 1873.

method of killing the small animals, destroyed the last of the game there.[1] By the late 1880s Gembrook was no longer of any use for acclimatisation purposes and the reserve was eventually sold by the Government; the ASV received £1,000 from the sale to recompense it for money spent there in the past.[2]

The ASV's involvement with fish acclimatisation was rather more successful. Although the Zoological Committee and the ASV had plans to import live fish ova from Britain, the early success in this area was organised from Tasmania. After several failed attempts, a number of tenacious Tasmanians eventually succeeded in transporting live Atlantic salmon and trout ova from Britain to the island colony in 1864. Some of the salmon ova were left in Melbourne where, much to the delight of the Council, they hatched and the young fish appeared to thrive after being released in the rivers. However, the success ended there; the salmon did not return to the rivers to spawn as they should have done. The experiments with salmon in Tasmania were no more successful and ultimately, they were abandoned in both colonies.[3]

The trout experiments were more successful. After various experiments with some success in the 1860s, systematic work to distribute trout around the colony began in 1871 under the direction of Albert le Souef. At first, many of the ova were acquired from the Tasmanian fisheries, but gradually the hatching ponds on Samuel Wilson's property at Ercildoune became the principal source. Wilson had spent some years unsuccessfully encouraging the introduction of salmon[4] and finally settled on producing trout. Le Souef and his staff placed the young fry in safe locations around the colony, usually late at night in order to avoid the heat of the day and to keep the area a secret from the greatest threat to the proliferation of the trout, the poacher.

The acclimatisation of trout was one of the ASV's few successes. It tried to introduce other fish, including carp, gouramier from Mauritius, English crabs and herring. In the case of carp, they succeeded only too well; the fish, which are bottom feeders, destroy the river and lake environments for the more delicate

---

1. Godfrey, p.16.
2. ZASV minutes, 3 February 1902, and *Forty-fifth annual report of the Zoological and Acclimatisation Society of Victoria*, 1909, p.11.
3. *Fourth annual report of the Acclimatisation Society of Victoria*, p.9; Walker, *Origins of the Tasmanian trout*, pp.32, 44; Godfrey, pp.18-19.
4. Wilson sponsored several consignments of the ova from Britain and California during the 1870s.

native fish.¹ The other fish did not prove successful; the ASV were either defeated by the difficulties in bringing them out to Australia or in finding them a suitable environment in which to live. The Zoo's direct involvement in pisciculture ended in about 1890 but the interest continued. In 1909, the Government built a fish hatchery in the Zoo for trout ova which were later destined for the ponds in Studley Park.

The third category of animals introduced by the ASV included those which were imported to enliven the natural Australian environment. Birds, usually referred to as song-birds, were the principal creatures in this category, described by McCoy as the source of 'those varied, touching, joyous, strains of Heaven-taught melody, which our earliest records show, have always done good to man.'² English thrushes, blackbirds, larks, starling and others were transported from Britain in cages.³ They were then given time to recover from the journey before being released at selected locations in Melbourne and around the colony. Many members of the Council objected to the inclusion of these apparently useless creatures on the ASV's list of desirable animals.⁴ However, the birds and other animals brought into the colony largely for ornament had influential supporters.

The most articulate supporters of this variety of acclimatisation were Edward Wilson and Frederick McCoy. In justifying their stance on such creatures as glow-worms and larks, they stressed the useful aspects of otherwise apparently useless animals. Wilson gave an eloquent description of the value of glow-worms: 'The young people of [Australia], like the young people of other lands, will fall in love with one another, and will indulge in their evening rambles; and, as in other lands, probably they will occasionally fall a little short of topics for conversation. What debt of gratitude would then, so circumstanced, not owe to anyone who should provide them with such materials as the light of the glow-worm and the song of the nightingale?'⁵ McCoy talked of the larks and other

---

1. Rolls, pp. 333-334; the European carp was declared noxious in 1962.
2. McCoy, 'Anniversary address', p.34.
3. McCoy later experimented with transporting the birds' eggs with the intention of hatching them in Australia (ASV minutes, 18 July 1865).
4. For example, William Lyall, on his resignation due to inability to attend meetings, complained that there were 'so many useless animals' included in the ASV's list (ASV minutes, 28 July 1863).
5. Wilson, *Acclimatisation*, p.15. The first batch of glow-worms was sent out by Wilson in 1860; they were described by Mueller as 'interesting and ornamental [and] may prove useful as enemies of the aphis'. (Mueller's report, October 1860, PROV, 1189, 748); in November they were transferred to the care of McCoy in the University.

song-birds as those 'delightful reminders of home' which were capable of 'sweetening the poor man's labours, inspiring the poet with the happiest thoughts, and softening and turning from evil even the veriest brute that ever made himself drunk or plotted ill against his neighbour.'[1]

Two successive Governors of the Colony had presented equally bizarre if charming arguments to support the introduction of particular animals. Sir Henry Barkly spoke in support of Wilson's belief in the value of acclimatising monkeys. Wilson, he said, was a thorough acclimatiser who supported 'the acclimatisation of monkeys as exceedingly useful, for the amusement of the wayfarer whom their gambols would delight as he lay under some gum tree in a forest on a sultry day.' He then added some fond memories of time spent in Guiana watching various monkeys and concluded, 'I am inclined to concur with him that it is desirable to acclimatise the monkey tribe if it can be done.'[2] The following year Sir Charles Darling stated that he did not like monkeys, but suggested the boa constrictor, which could 'be made one of the most interesting drawing-room pets possible.' He had seen them 'introduced suddenly amongst a party and made to rear their heads over a piano; and although a little alarm was at first created, the creature soon became an object of interest and curiosity.'[3] However, no attempt was made by the Council to introduce the boa constrictor for this purpose.

The attitude of Wilson, McCoy, Barkly and Darling to the value of particular animals was typical of the approach to nature in many popular natural history journals published in Britain in the first half of the nineteenth century. Under the influence of natural theology[4] all natural objects were regarded as having been created for a specific purpose, and much of the content of the journals was given to sometimes odd explanations of the precise use for which an object was originally designed. However, in Britain in the 1860s, such explanations were no longer being used in natural history journals as Darwin's theory of natural selection challenged the principles of natural theology.[5]

The conflict did not seem to alter Edward Wilson's attitude and he continued to send birds from Britain to Australia. In one instance, early in the ASV's

---

1. McCoy, 'Anniversary address', p.34.
2. *First annual report of the Acclimatisation Society of Victoria*, p.27.
3. *Second annual report of the Acclimatisation Society of Victoria*, pp. 32-33.
4. This doctrine was most authoritatively expressed in William Paley, *Natural Theology, or evidence of the existance and attributes of the Deity* (London, 1802).
5. Barber, pp.291-294.

The Utilitarian Zoo

history, the Council attempted to take control over the types of birds which he was sending from Britain. In October 1862, on discussing a list of items contained in a consignment he was preparing to forward, the Council expressed its preference for game rather than for the ornamental birds which it considered useless. It also asked Wilson specifically to exclude the starling because there were suggestions that it was 'not only useless but injurious.' Wilson responded by sending the starling accompanied by a letter in which he expressed his belief that it was 'a most useful friend to the cattle and as a grub destroyer ranking next to the rook and further that there was every reason to believe that its egg destroying habits had been largely exaggerated.' The Council deferred to Wilson's judgement because 'he was better capable of judging with all the means of information at his disposal in England.'[1]

So, throughout the 1860s song birds continued to arrive from Britain. The Council distributed them, some were liberated in the colony and some were sent to New Zealand and the Pacific islands on request. Some of the song birds adapted so well to the Victorian environment that they multiplied to pest proportions. Complaints about the sparrow destroying fruit crops began to arrive at the ASV's office. At first the Council took no notice; Thomas Black suggested that the complaints were being sent from 'a small number of persons [who] were not all of a trustworthy character,' adding that 'the English people were naturally given to grumbling, and not only the sparrow but everything else introduced by the Society would be found fault with by some.' Nevertheless the Council felt it necessary to investigate the sparrows. On researching the birds' diet, it found evidence to suggest that sparrows lived on beetles, grubs and flies rather than fruit. Nevertheless, the Council bowed to public pressure and agreed to support an amendment to the game laws. In 1871, the sparrow and another nuisance, the minah, were excluded from the protection of the Act to Protect Game.[2]

The ASV became involved with the game laws in the early 1860s. When a bill to provide for the protection of native and imported game was under consideration, several members of the Council examined it and suggested some alterations.[3] The Act, passed in 1862, gave full protection to all of the imported ani-

---

1. ASV minutes, 3 June 1862, 28 October 1862.
2. ASV minutes, 21 March 1871, 11 August 1868, 30 November 1869, 19 May 1871; *Victorian Hansard*, 23 November 1871.
3. ASV minutes, 27 November 1861; Council members involved in this process included Samuel Bindon and Hugh Chambers.

mals and seasonal protection to native animals. Although animals could be added to the protected list without recourse to Parliament, amendments were necessary to exclude or modify conditions concerning named animals.[1]

Once the law was in place the ASV Council worked at ensuring that it was enforced by sending a circular to suburban municipal councils advising them of the Act and asking them to advertise the provisions. The police were advised of the locations of animals being liberated and asked to assist with their protection. Despite these measures the animals and birds were never safe from destruction in or near the metropolitan areas. McCoy appealed to the women of the colony to remonstrate with children 'on the cruelty, folly, and selfishness of killing creatures', adding that, 'if the Ladies would do all this for us, the evil would disappear, and they would have the honour of assisting in a most material degree in accomplishing the great objects of the Acclimatisation Society.' The Council went as far as to form a bird protection society designed to appeal to young boys who were believed to be responsible for raiding nests.[2]

The ASV continued to monitor matters concerning the game laws. In 1871 and 1872 it recommended amendments to the Act which removed the sparrow, the minah and the hare from the complete protection of the law.[3] Some years later, it lobbied Government to put controls on the use of the swivel gun, largely in an effort to protect imported game. Later on, as the acclimatisation activities waned and the zoological collection became more important, the Council grew concerned about the protection of native animals. In 1891, for example, it recommended a closed season for the bustard and, the following year, a closed season for the kangaroo.[4] In 1895, after much persuasion, the Council agreed to support the Field Naturalists' Club in its attempts to have Wilson's Promontory declared as a permanent reserve. While the ASV's early interest in the game laws was certainly part of the acclimatisation process, it is likely that the later concern about native animals was significantly influenced by Albert le Souef and his son, Dudley, whose knowledge on the subject was considerable.

---

1. 'An Act to Provide for the Preservation of Imported Game and during the Breeding Season of Native Game', 24 Vict., No. 161.
2. ASV minutes, 2 July 1862, 29 July 1862, 30 December 1862; McCoy, 'Anniversary address', p.35; *Second annual report of the Acclimatisation Society of Victoria*, p.38.
3. Amendments to the game act discussed at ASV and ZASV meetings 1871-1872 with a special meeting on the protection of the hare on 29 July 1872.
4. ZASV minutes, 19 January 1891, 19 September 1892.

The Utilitarian Zoo

During the 1860s the ASV attempted to pursue several other objectives which it had included in its list in 1861. At the suggestion of Wilson, based on an idea he picked up from the *Société zoologique d'acclimatation*, the ASV organised dog shows in 1864 and 1865. Bindon and McCoy justified the exhibition by quoting the first objective of the ASV which referred to the 'perfection of domesticated animals' and claiming that an exhibition would aid 'the recognition and improvement of the various breeds.'[1] While the first dog show was successful, the novelty had worn off by the second exhibition, and that was the last dog show organised by the ASV.

In 1863 Wilson, who had returned to Victoria briefly, introduced monthly meetings in order to encourage discussions on zoological and acclimatisation topics amongst the ASV's members and their guests.[2] Papers on aspects of acclimatisation were read at these meetings and, prompted by Bindon, were published with the annual report.[3] The subjects which were discussed included silk culture, game birds, mulberry, fisheries and acclimatisation in general terms. This was Wilson's last significant contribution to the ASV. He attended his final ASV Council meeting in September, 1864 and left the colony shortly afterwards,[4] although he maintained close contact with the ASV from Britain.

In the years following Wilson's departure the ASV began to decay and, by 1870, the widespread interest in its activities which had been evident at the beginning of the decade had abated. Subscriptions, Government grants[5] and animal donations became smaller and fewer. The ASV had not yet had any major successes and morale was low. The conditions in the gardens in Royal Park were shabby with the main gates considerably dilapidated and the house, provided for the secretary, in a poor state of repair. Edward Wilson's silver grey rabbits had escaped and, with the help of some wallabies, had managed to destroy many of the flower beds. On top of all of this, the ASV's secretary had returned to Europe leaving the accounts in such disorder that the Council strongly suspected embezzlement or, at least, gross incompetence. The animals living in

---

1. ASV minutes, 12 January 1863.
2. ASV minutes, 29 December 1863.
3. ASV minutes, 15 November 1864.
4. He never returned to Victoria after this; before he left he resigned from the position of president to be succeeded by William Haines, the former Chief Secretary. His final gift to the ASV, which arrived days after he died, was a Wardian box of truffle mould from France which was given to William Guilfoyle, director of the Botanic Gardens (ASV minutes, 28 January 1878).
5. The financial situation is discussed in more detail in Chapter Three.

The Utilitarian Zoo

Royal Park in 1870 were representative of those which had comprised the collection in the late 1860s. They included a large herd of angora goats, miscellaneous deer, a variety of game birds, some cattle and sheep, a small number of native animals, and an odd collection of traditional zoo animals consisting of a zebra, a leopard and eight monkeys.[1]

Ten years of concentrated acclimatisation activities had demonstrated the relationship between expediency and the colonial zoo. Acclimatisation and the newly founded Zoo had been merged in 1858 when the Government's representatives ensured that public funds were spent on animals which were considered useful rather than entertaining. The foundation of the ASV, using the Zoological Committee as the basis of its administration, solidified this relationship. Although much of the evidence of this period vanished from the Zoo as the flocks and herds were sold, there were some lasting results on the Melbourne Zoo from the acclimatisation era. On a practical level, the administrative experience of locating, transporting and caring for these animals, as well as the network of connections, provided a useful basis from which to develop a zoo. In addition, two features which became important elements in the development of the Melbourne Zoo, were incorporated as a result of the ASV's activities in the 1860s. These were the establishment and layout of the grounds in Royal Park and the role of native fauna in the zoological collection, both of which will be discussed in the next chapter. Finally, the most obvious role of acclimatisation in the history of the Melbourne Zoo was to separate it from the Botanic Gardens. A distinct geographical location and an independent administration helped to keep alive the ideas of the Zoo's founding fathers of 1857 until Melbourne was prepared to support a zoological garden as a leisure centre.

---

1. ASV minutes, 27 May 1870.

3. The Australian Zoo

In 1870, when the Acclimatisation Society of Victoria decided to develop the zoological collection for public entertainment and leisure, the Melbourne Zoo entered on a decisive new phase in its history. It is sometimes suggested that the Council reluctantly assumed the role of zoo keeper. In her article on the history of the Acclimatisation Society, Linden Gillbank concludes: 'we should thank [the ASV Council members] for setting in motion the establishment of something they really didn't want - a zoo.'[1] There is, however, a deeper continuity than Gillbank allows between the initial determination to found a zoo in 1857 and the developments of the 1870s. Seen in this perspective, the acclimatisation activities of the 1860s may be seen as an episode, almost an interlude, in the history of the Zoo.

The key figure in the change of direction was Albert le Souef (1828-1902), Usher of the Black Rod in the Legislative Council of Victoria. He was appointed to the position of secretary to the ASV in 1870, a position which had previously been held by people who carried out the instructions of the Council as directed. Le Souef altered the nature of the position at his first meeting with the Council, when he presented a report which included suggestions and ideas couched in a way that made it difficult for the members present to argue. From then on le Souef took control of the ASV. He lived in a house at the Zoo and spent all of his spare time working in an honorary capacity for the ASV.[2] He raised his family there and three of his sons were to become influential administrators in Australia's zoos.

Albert le Souef had spent much of his youth in the Victorian bush, firstly with his father, who was protector of Aborigines on the Goulburn River, later as a bushman and station manager. His life in the 1840s and 1850s was eventful and he proved himself to be a resourceful and experienced bushman who was undaunted by bad luck. He virtually ignored the gold rushes going on around him; at one stage he planned to sell cattle to the gold diggers but, after visiting a gold field near Bendigo, he was so appalled by the living conditions and behav-

---

1. Gillbank, 'The origins of the Acclimatisation Society', p.269.
2. Le Souef continued to work in an honorary capacity until 1882 when he was given a small salary (ZASV minutes, 14 May 1882).

iour of the diggers that he resolved not to continue with that idea.[1] He settled down in the mid-1850s when he married Caroline Cotton, formerly of a station on the Goulburn river.[2] In 1863 he was appointed Usher of the Black Rod, a position he held until 1893. His work in the Melbourne Zoo allowed him to use the experience he gained in his youth and his undoubted love and knowledge of native animals was transferred to caring for and developing the collection in Royal Park.

His impact on the direction of the ASV was immediate. As soon as he took up the appointment he suggested that 'it would tend to popularise the Society very much if a portion of the grant recently made to the society were set apart for zoological purposes..In many large cities of Europe where zoological collections exist, they have always proved not only instructive but one of the most popular attractions.'[3] Le Souef's vision resurrected an aspect of the original plan of the Zoological Society of Victoria to which the ASV had attached little weight. The current Council[4] was supportive and, in 1872, it changed the name of the Society from the Acclimatisation Society of Victoria to the Zoological and Acclimatisation Society of Victoria (ZASV) to reflect the additional interests.[5]

The Council's attitude to recreational responsibilities had been erratic in the past. In the months following the foundation of the ASV, in 1861, the newly formed Council had been prepared to take over the public aspects of the ZSV as well as the acclimatisation activities. In accordance with this obligation, the plans for the base in Royal Park included features suitable for public recreation as well as provision of a depot for acclimatisation purposes. On announcing a competition for a design for the new gardens the Council had made it clear that it wanted 'a system of landscape gardening for zoological purposes.'[6] A design by Alfred Lynch, who had been engaged at the time in laying out the grounds of the Horticultural Society at Burnley, was chosen against six others,

---

1. A.A.C. le Souef, 'Personal recollections of early Victoria', p.48; for example, he had known the Porcupine Inn before the gold rushes as ' a comfortable, quiet country inn'; now 'every room in the house [was] crowded and beds made up on the floor and on the tables and the place was filled with half-drunken diggers. The meals were of the roughest sort and the prices exorbitant'.
2. Fourth daughter of the late John Cotton of Doogallook station, Goulburn river (le Souef, 'Personal recollections', p. 42].
3. ASV minutes, 3 June 1870.
4. Appendix VI, Council of the Zoological and Acclimatisation Society of Victoria, 1870.
5. ZASV minutes, 15 March 1872.
6. ASV minutes, 3 May 1861.

two of which were submitted from Mueller's department.[1] Facilities, including a refreshment tent and benches, had been added some months later; but for much of the decade, these were the only provisions made for the comfort of visitors.

Clearly the ASV Council members had not considered public entertainment to be a priority. The expense of keeping animals in the park, the constant problem of destructive children roaming unaccompanied through the grounds and, possibly, criticism in the British Journal, *The Field,* about Victorian's acclimatisation subscriptions being wasted on maintaining a menagerie, had discouraged the Council from promoting the entertainment value in the Royal Park depot. Evidence of public interest in a zoological collection was largely ignored (for example, the *Yeoman and Australian Acclimatiser,* 2 November 1862, suggested that: 'An attractive display of rare animals tends greatly to keep up and excite popular interest'); the ASV's combined policies of dispersing all animals as soon as they had recovered from the trip to Melbourne and of holding few native animals at Royal Park had done little to encourage visitors to the Park and their numbers fell considerably.[2]

The ASV had not been under financial pressure to provide the public with entertainment in the early years because the Government was generous and uncomplicated with its aid. However in 1864 the first signs that public funds might become more difficult to obtain in the future had emerged. Edward Wilson had warned that the 'legislature were greatly in ignorance of what the Society had done and was doing for the colony.'[3] He had introduced some activities, including the *conversazione,* to increase public awareness about the work. Even after he left the colony, he had continued to stress the importance of maintaining public interest. In 1865 he had written from England 'urging upon the Council the desirability of keeping as much as possible at the Society's gardens for the people to see.'[4] The ASV had not heeded his advice and, as Wilson pre-

---

1. ASV minutes, 29 May 1861; two other entrants were Humilis [?] and Omega; the competition was judged by a Mr Skene on behalf of the Government at the invitation of the ASV.
2. ASV minutes, 2 December 1862, 15 September 1863, 6 September 1864. Correspondence in the *Field* and the *Times,* reprinted in the *Yeoman and Australian Acclimatiser,* 13 September 1862. The problem with children who 'frequently molested' the animals was first mentioned in September 1863; the matter emerged at irregular intervals over the following years.
3. ASV minutes, 9 February 1864.
4. ASV minutes, 19 September 1865.

dicted, the Government had begun to make it clear that stringent conditions would be attached to ASV grants in the future.[1]

The one instance during the acclimatisation era when the ASV had seriously considered establishing a zoo was in 1866. The unstable financial position had been an important driving force behind the renewed interest in public entertainment. It was hoped that a zoological collection in Royal Park would raise revenue either by charging a fee to look at some exotic animals or by generating enough goodwill to encourage Members of Parliament to view the ASV's needs favourably. McCoy had begun the debate when he suggested that 'in every case a pair of each animal should be kept at the Royal Park.' (This may have had more to do with McCoy's interest in a supply of animals for the National Museum: two years later he asked for a pair of each quadruped and bird in the collection for the National Museum.) A week later a proposal to establish 'a complete zoological collection in conjunction with the ASV' had been considered just as a thorough investigation of the financial position got underway. The plan was to buy a small menagerie belonging to a Mr Stutt of Bourke Street. This collection, which included lions and bears, would have formed the nucleus on which the future collection could have been based. However, after much negotiation, the plan collapsed as several members of the Council rejected the proposal, ostensibly because the deal was not to their satisfaction. Yet there was no further negotiation with Stutt and the idea of establishing a zoo had been abandoned once again.[2]

Le Souef revived the idea on his arrival in 1870. He was motivated by two factors; an interest in continuing the acclimatisation activities despite the poor outlook, and a desire to harness the changing leisure demands of the public to the advantage of the ASV. As he transformed the Gardens in Royal Park over the next few years, the wisdom of his decision to set the ASV on a new course became more apparent.

The financial problems of 1866 had not abated by 1870 and le Souef was aware that acclimatisation alone would not sustain the ASV. Adverse criticism,

---

1. A proportion of the grant was to be matched by public subscriptions (ASV minutes, 24 January 1865).
2. ASV minutes, 5 June 1866, 12 June 1866, 26 August 1866, 7 May 1867, 28 May 1867 (decision on buying Stutt's animals: in favour, McCoy, Castelnau, R.B. Smyth and Levey; against, Champ, Dobson, ARC Selwyn, Lang, Purchas, Venables and Steavenson), 15 September 1868. Stutt is not mentioned by St Leon, op.cit., in connection with any circus.

particularly in Parliament, was damaging to the cause and, in the early 1870s, there were several such comments. For example, when the ASV grant was under discussion in 1870, a Member of the Legislative Assembly claimed that 'there was a growing feeling that the efforts of the Acclimatisation Society had not always been successful, and, as regarded the introduction of some animals, absolutely injurious.'[1] Three years later another Member of the Legislative Assembly declared that the ASV was 'hitherto not of the slightest use to the country, while they were practically responsible for the introduction of the rabbit and the sparrow.[2]

Yet public opinion was by no means uniformly hostile. In 1874, a member of Parliament made reference to the valuable work of the ASV in stocking the woods and forests with game 'so that the colonists could indulge in field sports,' adding, 'it was a fact that there was no inducement for people of money and leisure to remain as residents of the colony.' He went so far as to suggest that the ASV should be granted additional funds 'to enable it to relieve the country from being a barren waste and supplant such vermin as the wallaby and the native cat with valuable game.'[3] The newspapers, most particularly the *Argus* and the *Australasian*, also expressed continued interest in acclimatisation activities. With the criticism being balanced by the favourable opinions, the continuation or otherwise of acclimatisation was in the hands of the ASV. With le Souef in charge, the activities were going to be maintained for the present.

The second major factor which influenced le Souef was a noticeable change in the leisure patterns of the people of Melbourne. There was striking evidence of this in the painful debate about the direction of the Botanic Gardens in the early 1870s. The gardens which Mueller had created were scientific and educational, with a view to the economic requirements of a developing colony. But to the public they looked 'dull and unplanned' and not the 'continuous floral displays, sweeping lawns and vistas' which were increasingly expected.[4] The Botanic Gardens had been the subject of complaints for some years; the *Argus*, for example, commented on 'the lack of 'beauty' despite the large amounts of Government money which had been spent on it. The *Argus*, like many other newspapers, attacked Mueller personally: 'he is a most difficult person to deal with.' It

---

1. George Higinbotham in *Victorian Hansard*, 22 November, 1870.
2. L.L. Smith in *Victorian Hansard*, 12 June 1873.
3. John Curtain, member for North Melbourne, in *Victorian Hansard*, 24 November 1874.
4. Pescott, *The Royal Botanic Gardens, Melbourne*, p.79.

cited his 'unlovely qualities', including 'his undignified subserviency to all men who possess, or are supposed to possess, power or influence'.[1] On the other hand, there was universal agreement in the colony that Mueller was a highly accomplished botanist. A Board of Inquiry investigated the situation and came out in support of popular public opinion. The Board recommended that while Mueller should remain in charge of the scientific branch of the operation, a curator would manage the gardens with responsibility for their 'proper design and maintenance.'[2] However distinguished his scientific accomplishments may have been, it was clear that Mueller was not the man to provide the public with the pleasure grounds that they demanded.

The problems experienced by Mueller at the Botanic Gardens provided an insight into, and a warning of, changing public interests. Criticism of the lack of public facilities or the appearance of the ASV Garden was more muted, perhaps because expectations were lower. There had been some complaints before the arrival of le Souef that the Gardens were not attractive or well managed.[3] Such comments were not widespread as the ASV grounds were well known only to the people living in the immediate neighbourhoods of Carlton, Brunswick and Hotham.[4] Nevertheless, by developing the ASVs recreational facilities, le Souef pre-empted the kind of criticism which led to Mueller's downfall.

He began to implement his plans immediately. His approach to developing the new institution took the poor financial situation as well as the acclimatisation commitments of the ZASV into account. Unlike the two previous occasions

---

1. *Argus*, 15 July 1872; and J.M. Powell, 'A baron under siege', pp.18-35.
2. 'Official report of the Board of Inquiry upon the botanical Gardens' quoted in Pescott, *The Royal Botanic Gardens*, pp. 82-83. Also in May 1873 Mueller resigned from the ZASV Council. The ZASV had changed the tenure of office holders in the Council from perpetual to annual. Mueller had held the position of Vice-president since 1861. He was reelected to this position in the first year of the new system, 1872-1873 but on completion of that year he failed to turn up at meetings, apparently in the mistaken belief that he was no longer a member of Council. Despite two letters assuring him that he was still on the Council, he resigned. He was not awarded a medal or accorded a speech of thanks. Thomas Black, who had handed over the presidency at the same time, was offered a gold medal (the only one ever mentioned by the Council) but he refused it, preferring lengthy speeches of thanks instead (ZASV minutes, 24 March, 1873). Nineteen years later, Mueller was elected an honorary member of the ZASV 'as the Baron was not only an eminent scientist, but he had on many occasions rendered special services to the Society' (ZASV minutes, 5 September 1892).
3. Godfrey referring to the era before the arrival of le Souef (*Victorian Hansard*, 24 November 1874).
4. Reference to these suburbs appeared occasionally, for example, *Fifth annual report of the Acclimatisation Society of Victoria*, p.10, and the *Age*, 23 November 1874.

in 1858 and 1866, when it was proposed to develop a zoological collection, there was no suggestion of buying a ready-made menagerie of foreign animals with keepers and cages. Instead, he took a radical approach to zoo development by concentrating on creating a pleasant, landscaped garden and filling it with Australian animals.

Le Souef's decision to establish a zoo based almost entirely on native fauna was contrary to contemporary trends in Europe, of which he had second-hand knowledge. The principal function of European zoos as they evolved during the nineteenth century was to provide the audience with a glimpse of exotic creatures from apparently inaccessible regions of the earth. Indigenous animals, no matter how interesting to a zoologist, could not attract a local audience. Consequently, large vertebrates from Africa such as the elephant, the rhinoceros, the lion and the hippopotamus formed the basis of the most popular zoos.[1] Australian fauna, despite being from a distant land, never matched the popularity of the African animals in Europe although most zoos held a kangaroo, an emu, an opossum and some Australian birds.[2]

Le Souef was aware that a zoo comprised principally of Australian animals would not satisfy his Victorian audience, but he had several good reasons to adhere to his policy throughout the 1870s. The most obvious reasons were dictated by the financial condition of the ZASV. The local animals were easy to acquire, cheap to maintain (being principally herbivorous), and they were useful for exchange with zoos and acclimatisation societies abroad. But it is likely that he also found the idea of an Australian zoo appealing, believing that the fauna 'would yet in itself prove most interesting.'[3] and, on several occasions, he stressed the need to have as complete a collection of native fauna as possible.

The collection of Australian animals in the Melbourne Zoo grew rapidly. Most of the animals were donated by people living in the colony. By 1875 it included Tasmanian tigers, Tasmanian devils, a koala, wombats, various wallabies and kangaroos, a number of opossums, some kangaroo rats and many Australian birds. These lived alongside the deer and other remnants of the acclimatisation era. Le Souef also purchased two African lions and some American black bears because they had been available in Melbourne at a cheap price. The lions and

---

1. Mullan & Marvin, *Zoo culture*, p.74.
2. The ASV and its successor, the ZASV, kept many of the European zoos supplied with Australian animals; see successive annual reports 1860-1890.
3. ASV minutes, 3 June 1870.

bears, with the monkeys which had been living in the Zoo for some time, were the only concessions he made to the public's demand for the exotic.[1]

The unusual approach to the development of the zoological collection received some public support. The *Age* (23 November 1874), for example, agreed that a representative collection of the fauna of Australia was a more pressing need than the acquisition of 'zoological luxuries,' such as elephants and tigers, when funds were low. In 1876 'The Vagabond' advised a similar approach: 'First obtain a complete collection of every Australian animal and bird; fishes and an aquarium you might add in time. But let the gardens be an Australian menagerie, where the rising generation may learn what living things surround them on this continent' (*Argus*, 2 December 1876). Eight years later he complimented the Council on the collection of native fauna: 'Boys who have never seen the joys of nature which make up a juvenile poacher's experiences, or know not the glories of animal life on a Riverina plain, may learn lessons of the highest importance by studying the habits of the marsupials, the birds, and the reptiles of his native land...our youngsters will learn all this in the Zoological-gardens' (*Argus*, 21 July 1884). As anticipated, however, the majority of local visitors clustered around the few imported animals and paid scant attention to the indigenous collection.[2] Le Souef was undaunted; he maintained his course, declaring on more than one occasion that he was not going to increase the international collection until the ZASV's financial position was more stable.

Improving the appearance of the gardens was the other major project which le Souef could work on without needing access to large monetary resources. He believed that an attractive garden should encourage public support for the long term development of the Zoo.[3] The ASV property in Royal Park covered an area of 40 acres, which was a relatively generous size for an urban zoo.[4] The gardens had originally been laid out in the early 1860s by the ASV to a plan designed by Alfred Lynch. Since then, some trees, shrubs and flowers had been planted. But the Council's lack of interest in promoting the recreational aspect of its activities meant that there had been little concerted effort to maintain the appearance of the grounds.

---

1. *Proceedings of the Zoological and Acclimatisation Society of Victoria*, Vol. IV, 1875, p.29.
2. *Argus*, 2 December 1876, and other contemporary newspaper reports.
3. ZASV minutes, 6 September 1872.
4. The London Zoo was in an area of about 25 acres in 1834; this was gradually increased to 36 acres by 1976 (Blunt, *The ark in the park*, p.26); Adelaide Zoo, the smallest of the Australian Zoos, only covers 8 hectares, c.16 acres (*Adelaide Zoo*, 1984).

The Australian Zoo

Shortly after he arrived, Le Souef began to tidy up the gardens with the aid of a small gardening staff. The ground was ploughed and trees already growing in a nursery in Royal Park were transplanted.[1] Mueller was generous with his assistance, donating various pot plants on several occasions from the Botanic Gardens.[2] He also donated some water plants for the fish ponds which were, by then, serving both a useful and an ornamental function.[3] Pipes were laid to direct water to the higher parts of the grounds from the existing supply in the lower sections.[4] Within two years the gardens were under control and le Souef was ready to develop them. Over the following years, new paths were set out and previously inaccessible parts of the grounds were opened up to the public.[5]

The first map produced by the ZASV of the Zoo was published in 1875 after le Souef had begun to develop the gardens. Consequently, it is difficult to ascertain how much of Alfred Lynch's original plan had been executed by 1870 and how much of the layout in 1875 was in fact le Souef's design. Certainly, the ponds on the north-east of the grounds were made in the early 1860s,[6] and it is probable that the wide central and peripheral paths had also been laid at that stage. There is no evidence in the Zoo's records to suggest that Alfred Lynch's design of 1861 was resurrected in 1870. It is likely that the additional walks were designed entirely by le Souef; neither Mueller nor his successor, William Guilfoyle, gave any formal advice on the matter.

From 1874 the appearance of the gardens began to feature prominently in newspaper articles about the ZASV. 'Sketcher' in the *Australasian*, commenting favourably on the flower beds, the trees and shrubs, considered that 'inanimate nature...does its part well in administering to the pleasure of the Society's visitors' (*Australasian*, 12 December 1874). Later that summer a reporter from the *Age* recalled 'a delightful afternoon' spent among the trees and flowering shrubs and complimented the ZASV on the arrangement and management of the gardens (*Age*, 23 November 1874). The following summer the *Argus* declared that the flower beds and borders were 'as gay with rich and varied colour as was the

---

1. ASV minutes, 26 May 1870; the names of the trees were not specified.
2. On one occasion Mueller sent 150 unspecified potted plants to the Zoo (ASV minutes, 26 August 1870).
3. ASV minutes, 21 April 1871; water plants not specified.
4. ASV minutes, 15 January 1872.
5. *Argus*, 25 December 1875; 'Report of the Council' in *Proceedings of the Zoological and Acclimatisation of Victoria*, Vol. III, (Melbourne, 1874), p.9.
6. *Yeoman and Australian Acclimatiser*, 25 September 1863.

Flemington lawn on Cup day' (*Argus*, 25 December 1875). By the early 1880s the gardens were established. A social writer for the *Herald* described them enthusiastically: 'the beds of brilliantly tinted blossoms... lent a charm of freshness and beauty to the scene while a hundred blended perfumes wafted on the breeze, and seemed to make the air heavy and languid with their sweetness' (*Herald*, 23 January 1882).

By 1880, the Zoo's landscape had shifted from resembling a farm to being an attractive recreational garden. Nevertheless the influence of the acclimatisation era was apparent in the unusual character of the Zoo grounds. Instead of small cages lining paths, the Melbourne Zoo was dominated by large paddocks which contained the kangaroos, wallabies, emus, cassowaries, as well as deer and other herbivorous animals. Smaller native animals were kept in reasonably spacious paddocks with special features; the koalas, for example, had a paddock with gum trees, and the wallabies were given a heap of stones. The areas in between these enclosures were planted with trees, shrubs and flowers. Aviaries, ponds and the small cages for the imported animals were scattered throughout the Gardens. Even as the number of imported animals increased in the early 1880s, their cages continued to be insignificant in relation to the size of the large paddocks. Eventually, however, le Souef decided to insert divisions in some of the larger paddocks in order to assist visitors in identifying the varieties of species which were well represented in the Zoo. The kangaroos were the first to be split up to enable visitors to see the different gradations from the great grey or red kangaroo, both about six feet high, to the small kangaroo mouse.[1]

In 1872 le Souef claimed that 'the zoological gardens bid fair to become as popular here as they are in all the large cities of Europe where they are established.' The response from the public during the 1870s justified that claim. Although there is no reliable way of ascertaining the figures, the numbers of visitors undoubtedly increased.[2] Some contemporary newspaper reporters described the public use of the gardens. *Liber* in the *Age* listed some general categories of visitors and their activities: 'elderly folks go there for an hour's quiet enjoyment; married couples with their progeny wander from the lions' den to the monkey house...lovers stroll about the garden arm in arm, whispering to one another the

---

1. *Argus*, 24 September 1881; ZASV minutes, 3 October 1881, 6 December 1886.
2. Le Souef estimated that there were 2500 visitors on a Sunday in November 1873 (ZASV minutes, 17 November 1873), and 5000 visitors on new years eve, 1878 ('Annual report of the Council', in the *Argus*, 25 February, 1879).

sweet nonsense that is usually uttered on those occasions...children go there for a romp...In fact people of all ages and all classes take delight in visiting that place where instruction may be found by those who seek it, and those who do not may pass away a couple of hours pleasantly enough among the green trees and flowering shrubs' (*Age*, 23 January 1875). 'The Vagabond' was more specific in his description of 'the usual heterogeneous Victorian crowd' in the gardens: 'Working men with their wives and children, tradespeople, clerks, schoolboys, sewing machine and show girls, with a touch of the larrikin element...I was struck by what I am told is true colonial style -the fact of the husband so often walking holding his children by the hand, and the wife pacing in front, as if she did not belong to them.' He added that the people appeared to be enjoying themselves: 'after going the round of the gardens they would sit or lie in the shade of the hedges of shrubs and revel in that *dolce far niente* which none but those who work six days in the week can fully appreciate' (*Argus*, 2 December 1876).

However along with the increase in the number of visitors came the problems of having to control what le Souef referred to as 'the ever growing larrikin evil.'[1] Since the mid 1870s the larrikins had been causing trouble in the Gardens by roaming through the ground in large numbers annoying the visitors, being abusive to the staff and disturbing the animals. Some of the more extreme acts of vandalism included setting dogs free amongst the animals and assaulting the head gardener.[2] However, it was the constant need to control unruly behaviour which exhausted le Souef and his staff. Their only defence at that stage was to post by-laws at the entrance gate and to get additional assistance from the police to ensure that the larrikins adhered to the prescribed standard of behaviour. The alternative, of course, was to restrict access to the Zoo by charging an admission fee; not only would that help to control the problem of unruly youths, but it would strengthen the Zoo's financial position.

The overriding issue of the 1870s was the Zoo's financial status. So far, Government grants and subscription fees had provided its only income, and both were unreliable sources. The issue was first discussed openly in 1874 when the Government decided to end its financial support of the ZASV completely. As a counter attack, the Council threatened to introduce an admission fee without permission in the new year. In anticipation of this move, it delayed exhibiting a camel which had been on the Burke and Wills expedition in 1860. The

---

1. ZASV minutes, 30 August 1880.
2. ZASV minutes, 11 May 1874, 25 April, 1875.

camel had been found in the Murray Downs and brought back to Royal Park. The Council, aware of the 'melancholy interest' attached to the animal and the enormous public demand to view it, withheld it with the intention of making it a star attraction later on. At the last minute the Government retreated and promised to continue supporting the ZASV. The plans to introduce an admission fee were dropped for the present. The camel went on exhibition but, shortly afterwards, it succumbed to the cold Melbourne winter and died.[1]

The admission fee issue was revived again in 1877. The growing popularity of the Gardens, and the ambition of the Council to develop them further, continued to place great strains on their financial resources.[2] This time le Souef and the Council had decided that an entrance charge was essential to the future of the Zoo and they were determined to resolve the matter permanently. During the 1870s the ZASV's income averaged between £2,000 and £3,000, some of which came from subscriptions and sale of stock.[3] Given the financial limitations, the Council was reluctant to import any of the expensive animals which were most popular in zoos abroad. European zoos in the same period had incomes of around £10,000 gathered from entrance fees and subscriptions.[4] Le Souef and the Council believed that an admission charge would stabilise the Zoo's financial position. But the introduction of such a plan was not at all straightforward.

The agreement by which the ZASV had tenancy of the grounds in Royal Park did not allow a charge to be levied at the entrance gates. Only the Government could alter this arrangement and, for a variety of reasons, successive Governments were not prepared to do this. The main obstacle to restricting access to paying visitors only was the formidable Chief Secretary, Graham Berry, who let it be known that he objected to an entrance charge. A self-educated man of humble origins, he has been described by Serle as one who worked for 'the relief of hardship, and the material, educational and moral elevation of the working man' and was 'idolised by the masses and hated more by the upper classes than any radical in Victorian history.'[5] For Berry, the question of admis-

---

1. 'Report of the Council', in *Proceedings of the Zoological and Acclimatisation Society of Victoria*. Vol. IV, p.10; ZASV minutes, 17 August 1874, 28 June 1875.
2. 'Report of the council', in the *Argus*, 26 February 1878.
3. Financial statements in the annual reports of the Council, 1871-1880.
4. A.A.C. le Souef's report on his return from Europe, ZASV minutes, 30 August 1880.
5. Serle, *The rush to be rich*, p. 20.

sion fees was certainly a class issue. 'He would be extremely loth,' he told le Souef at one stage, 'to interfere with the opportunity at present enjoyed by the people of visiting the gardens on Sunday, as they were shut out from almost every other place of recreation on that day' (*Argus*, 3 August 1883).

For several years Berry blocked the alteration of the tenancy agreement. As if by way of compensation, he agreed to a slight increase in the Government grant, and characteristically suggested that 'a fashionable day' should be inaugurated so that a high fee might be charged on one day a week. A band and other entertainments might be made available on that day to attract a crowd. The Council initially suggested making Saturday or Sunday the 'fashionable' day, but Berry refused to sanction that proposal. Eventually Friday was chosen with great reluctance, and the Council had little hope that the new system would work.[1]

Some months later there was a change of Government and the new Chief Secretary, Sir Brian O'Loghlen, agreed to permit charges from Monday to Friday, but Saturday and Sunday were to remain entirely free.[2] This le Souef still considered inadequate and even potentially dangerous: 'If the gardens are made more attractive, as they of course will be under the new system, the people will come in such large numbers on Sundays as to be quite beyond control and something serious may happen in consequence, then again on Saturdays when children are not at school they visit the gardens in large numbers and seem to delight in doing all the mischief they can, the animals are petted and stoned and nearly driven mad and the buildings and cages bespattered with mud and dirt and unless there was a keeper for every cage it is impossible to put a stop to it.' He was so dissatisfied with O'Loghlen's deal that he recommended that the Council should borrow money from a bank, buy a site, transfer the zoological collection to the new location and leave the ZASV property in the hands of the Trustees of Royal Park. The idea was not treated seriously by Council. In October, O'Loghlen quietly conceded a Saturday charge.[3] Now only Sunday was free but le Souef had not given up the fight.

His next move was to organise a display of public support for a Sunday fee. A section of the community had always demonstrated its readiness to pay for admission to the gardens. In the early 1870s, as the new direction of the Zoo became known, the list of subscribers had grown considerably. In 1871 when the

---

1. ZASV minutes, 13 December 1880, 10 January 1881.
2. ZASV minutes, 22 August 1881. O'Loghlen replaced Berry in July 1881.
3. ZASV minutes, 22 August 1881, 17 October 1881.

subscription stood at two guineas, the income from this source was just over £99; this rose to £491 in 1872 when the levy was dropped to one guinea.[1] However, there were no privileges for the subscriber except 'the barren honour of seeing his name in the pamphlet of the Society's proceedings' (*Age*, 23 November 1874). Consequently, the ZASV could not sustain the interest of subscribers and, by the end of the decade, the number had dropped considerably.[2] The public's willingness to pay was demonstrated yet again when le Souef organised a petition over the Christmas period in 1881 seeking support for a Sunday charge. The public response was positive; nearly five thousand people, mostly male, supported the plea. On le Souef's instructions, only those who read the public notice about the petition were to sign their names; in addition women were to be discouraged and children were to be prevented from signing.[3] The petition was presented to the Chief Secretary, but to no avail.

Several times over the following years le Souef and the Council renewed their attempts to introduce a Sunday entrance charge. In 1883 they even tried to use the controversial move by Sabbatarians to keep such institutions as the Public Library, Museum and Gallery closed on Sundays to their advantage. 'If it was sinful to visit the gardens on Sunday,' one member of the Council argued, 'it would be better to have a charge, as then the number visiting the gardens on that day would be reduced by 50 per cent !' (*Argus*, 23 October 1884). This was a dangerous tactic. One of the reasons why *The Lord's Day Observance Society* did not demand that the zoo should also be closed was precisely because there was no charge.[4] When the Sabbatarians got their way with the Library and Museum, le Souef again tried to restrict public access to the Zoo on Sundays by suggesting that perhaps it should also be closed to all but members of the ZASV. He was most concerned about the possibility of a crush as the Zoo was now one of the few recreational venues open on a Sunday.[5]

---

1. 'Report of the Council' in *Proceedings of the Zoological and Acclimatisation Society of Victoria*, Vol. II, 1873, p. 7.
2. The ZASV never achieved the same number of subscribers again although, towards the end of the 1880s, the number of subscribers rose again; in 1887, for example, £336 was collected ('Report of the Council' in *Twenty-fourth annual report of the Zoological and Acclimatisation Society of Victoria*, 1888, p.9).
3. ZASV minutes, 9 January 1882.
4. Serle, *The rush to be rich*, p.158; in 1890 the Lord's Day Observance Society tried to persuade the Council to make Saturday the free day and close the gardens on Sunday; nothing came of it.
5. ZASV minutes, 9 July 1883. The Library and the Museum remained closed on Sundays until 1904.

The Australian Zoo

His concerns were realised as the number of Sunday visitors grew. On particularly popular days le Souef and his staff barely managed to keep order. In the late 1870s, five thousand Sunday visitors was regarded as a large number; towards the end of the 1880s le Souef estimated that crowds of thirty to forty thousand people were visiting the gardens on 'hospital Sunday.' However, Graham Berry, again Chief Secretary in 1883, in curious alliance with the Sabbatarians, was adamant a Sunday charge would not be allowed and offered a larger police presence instead. The Zoo remained free to the public on Sundays for the rest of the century and, as compensation, the Government continued the annual grant. Early in the twentieth century a small charge was introduced and, although it reduced the number of Sunday visitors, there was little protest.[1]

Despite the Sunday chaos, the combination of a Government grant, entrance charges from Monday to Saturday, subscriptions and other miscellaneous sources of income had placed the Zoo on a reasonably stable footing by 1881. Le Souef's agenda for the 1870s had been successful; the slow development of the Zoo using native animals and careful landscaping as the focus had allowed the existing operations to continue while the new activities were incorporated into an evolving structure. He was now confident that he could begin to develop the Australian Zoo into a zoo of international standards. The Council members had supported him throughout the decade, indicating that they were prepared to take responsibility for a recreational function along with their remaining utilitarian functions.[2] During the 1870s their essentially private organisation had become the management of a permanent metropolitan institution.

Le Souef's unconventional approach to establishing a zoo left the Melbourne gardens with some unique characteristics. The native animals, although initially not popular with the local people, became an important part of the collection. As the foreign zoological collection grew, le Souef maintained the number of Australian animals. His son, Dudley, who succeeded him as director in 1902, also believed in the importance of the role of Australian fauna in the Zoo and consolidated its position amongst the more traditional zoo animals. The other unique characteristic which emerged during the 1870s was the spacious layout of the grounds, the result of an unusual emphasis on herbivorous animals.

---

1. ZASV minutes, 6 May 1878, 6 August 1883, 29 October 1888; 'Report of the Council' in *Forty-eighth annual report of the Zoological and Acclimatisation Society of Victoria*, 1912, p.9.
2. At the end of the 1870s, the Council was still managing the reserve in Gembrook, monitoring its stock on properties around the colony and working on the acclimatisation of fish.

Although modifications were made as the collection grew and more cages were erected, the gardens remained uncluttered, with pleasant walks and generous enclosures. These characteristics, products both of necessity and deliberate policy, are still prominent features in the Melbourne Zoo.

4. The International Zoo

Le Souef's emphasis on the Australian zoo by no means precluded a second objective: to give the city of Melbourne a zoo of international stature. The task cannot have been simple. Limited resources presented the greatest obstacle to achieving their objective. They were also a long way from the centre of zoo development in Europe and were thus in danger of introducing outmoded features which had proved to be of no interest to a large audience. In addition, the pressure of being popular in order to be financially viable was not something any member of the Zoo's administration had had to contend with before. By 1900 the Zoo was of international stature, but the questions remain as to how that was achieved; what influenced the development, how did the administration cope with the limited resources and how did the Melbourne Zoo compare with the zoos of Europe?

The fourth phase in the Zoo's history was again largely orchestrated by le Souef. He had delayed the Zoo's move into its international phase throughout the 1870s despite tempting opportunities to break that policy. But his determination that no expensive, imported animals should be purchased or even kept in the gardens until the Zoo was on a sound financial footing held firm, and several chances of purchasing exotic animals were passed over. A circus offered to sell two elephants, hyænas and other exotic animals to the Zoo in 1877 but was refused. It later tried to exchange the same animals for some of the Zoo's stock and was again rejected.[1] But le Souef was confident that this situation would not continue indefinitely, and he began to prepare for a rapid expansion in the collection just as soon as the admission fees were introduced.

Central to his preparation was a trip he made to Europe in 1880 with the support of the Zoo. It had been his ambition to go there for some years to examine the latest developments and to study the management of other zoos. The housing and treatment of some of the more popular animals were of particular interest to him; how, for example, did zoos care for large African animals captured in the wild and transported to a small urban cage? He wanted to get some first hand information and advice about this and other matters. He also wanted

---

1. ZASV minutes, 29 January 1877, 28 January 1878.

The International Zoo

to establish contact with zoo administrators to facilitate future exchange arrangements. When the proposed tour was finally sanctioned by the Council, le Souef left for Europe taking with him money bequeathed to the ZASV by Edward Wilson and, for exchange purposes, some Australian animals, including kangaroos, emus, dingo pups and wombats.[1]

This was his first visit to Europe since he left England for Australia at the age of twelve. His wife, Caroline, accompanied, him making sketches of buildings and other features in all of the zoos. While he was away his eldest son, Dudley, who was employed as the assistant director, took on his duties. Dudley le Souef had begun helping his father around the zoo as soon as he had finished his education in England. Then, in 1874 at the age of nineteen, the Council had agreed to employ him formally. As well as doing routine duties, he had learnt taxidermy and built up a collection of stuffed animals for the Zoo's small museum. By 1880 he was fully capable of assuming all of the director's duties during his father's absence.[2]

Albert le Souef began his tour in London, where he exchanged the Australian animals for bears, badgers and other stock worth about £80. He then went to the Continent where he visited zoos in Amsterdam, the Hague, Rotterdam, Paris and Cologne. The opportunity to exchange information with his peers lived up to his expectations. The zoos' administrators had no reservations about discussing the internal workings of their respective institutions and they were most forthcoming on the subject which dominated the operations of the ZASV, zoo funding. The Jardin des Plantes in Paris was the only zoo free to the public, but it got an enormous grant from the French Government. Even with that, le Souef considered the animals 'poorly housed for such a city as Paris' and thought that 'the place had a neglected and uncared for appearance.' He was impressed with the zoos in Cologne and Rotterdam, 'being more what I should like to see our gardens at the Royal Park become.' Fund raising schemes which he learnt about in Europe included those based on the international sale and exchange of surplus animals. This trade appeared to be the source of considerable profit in European gardens, with game and other birds commanding particularly good

---

1. ZASV minutes, 6 October 1876, 27 September 1879.
2. A.A.C. le Souef, 'Personal recollections', p.17. Le Souef brought Caroline on all of the trips which he made for the Zoo; unfortunately her sketch books, which probably accompanied her on each of the journeys, are not in the Zoo archive. ZASV minutes, 28 September 1875, 2 September 1895.

The International Zoo

prices. The trip made le Souef all the more determined to pursue permission to have a charge at the Zoo gates. He returned to Australia confident that, with the knowledge he now possessed, the Melbourne gardens 'would rival some of those in Europe.'[1]

This was the first of many trips made by le Souef or one of his sons, Dudley and Ernest, over the next twenty years on behalf of the Zoo. The objectives of each trip were the same: to collect animals, make contacts, study the administration of the other zoos and take note of changes and developments in design, layout and facilities. They went to Calcutta, South Africa, America, and Singapore as well as to Europe to exchange and purchase animals. They also went to Tasmania and Queensland to collect native fauna. Some unsuccessful attempts to purchase animals by mail had cost the ZASV a substantial amount of money, so choosing animals in good condition and ensuring their safe transportation back to Melbourne were the major justification for these trips.[2]

To le Souef, studying international trends in zoo development was obviously an important reason for travelling. In the second half of the nineteenth century many new zoos had been founded on continental Europe, particularly in Germany. Their organisation and objectives were slightly different from that associated with the London Zoo model, the essential difference being that they were founded by shareholders and were entirely commercial operations. Consequently they were designed to attract as many visitors as possible and to encourage them to return regularly. Elaborate enclosures, catering facilities and function rooms as well as collections of exotic animals were amongst the most prominent features[3] although the focus of the attention undoubtedly remained with the collection of exotic animals. Such was the vogue that, in a twenty year period, one dealer alone, Carl Hagenbeck,[4] imported 1000 lions, 400 tigers, 800

---

1. ZASV minutes, 30 August 1880.
2. ZASV minutes, 18 January 1886, 16 December 1889. (Ernest, le Souef's second son, joined the Zoo staff in 1889, firstly as the accountant and, after further study, as the veterinary doctor.)
3. Mullan & Marvin, *Zoo culture*, p.111.
4. Carl Hagenbeck was perhaps, the best known zoo director in Europe in the late nineteenth and early twentieth centuries; he was renowned for his innovative enclosure design and is remembered for his work on the concept of the 'barless' zoo (Hancocks, *Animals and architecture*, pp.126-127).

The International Zoo

hyænas, 300 elephants, tens of thousands of monkeys, 150 giraffes, 51 rhinoceros as well as many others, all taken from their natural habitat.[1]

The greatest obstacle facing the Melbourne Zoo on buying animals in Europe was the price. A pair of zebras cost Dudley le Souef £170 in Paris in 1886, and that was out of a total budget of £300 for purchasing animals on that trip.[2] When Europe suffered a cold winter, as it did prior to Ernest le Souef's trip in 1891, many of the animals which had been held in outdoor enclosures perished. Of course, prices of animals soared as the wealthier zoos competed to restock their depleted collections and Ernest le Souef could not compete. He still managed to buy reindeer, wild barbary sheep[3] and, the prize of this journey, an American Bison for £145. Unfortunately, on the voyage back to Australia, the bison reacted poorly to the change in weather conditions when the ship left the tropics and, despite the efforts of Ernest le Souef and the ship's doctor, it died of inflammation of the lungs.[4] Such were the hazards of transporting animals around the world and it made the ZASV's task even more difficult. The Australian fauna brought to Europe was some compensation for the Melbourne Zoo's lack of funds. Animals which cost £20 in Australia could be exchanged abroad for about £200 worth of stock. Unquestionably, the most valuable native animal was the thylacine; on one occasion one of these marsupials, purchased locally for £12, fetched stock worth £90 on exchange. By way of comparison, a pair of kangaroos was swapped for a yak worth £40.[5]

The choice of foreign animals for the Melbourne collection reflected European priorities. Despite the policy of emphasising Australian fauna during the 1870s, the Zoo's administrators had purchased a few of the cheaper and more readily available foreign animals, including lions, tigers, various primates and bears, from sources in the colony and from existing exchange contacts. In many cases they were acquired from the large American circuses which toured the

---

1. Herman Reichenbach, 'Carl Hagenbeck's Tierpark and modern zoological gardens', *Journal of the Society of the Bibliography of Natural History*, IX (4), pp.573-585, quoted by Mullan & Marvin, p,112.
2. ZASV minutes, 2 August 1886.
3. A herd of barbary sheep in a zoo was described by Henry Scherrin as 'a good show' which breed well in confinement (Scherrin, *The Zoological Society of London*, p.202).
4. ZASV minutes, 31 August 91.
5. Dudley le Souef's report to the Council on his return from Europe, ZASV minutes, 2 August 1886.

continent from 1873 to 1891.¹ Although these animals were very popular with the local visitors, le Souef and the Council did little to promote them.

This attitude changed as soon as permission to charge an entrance fee was granted. Le Souef made it clear that popular attractions were absolutely necessary, and he quickly identified some of the animals which the Zoo should acquire as soon as possible. The elephant was at the top of the list, not because of curiosity value, as most Melburnians would have seen one when the big circuses visited the city, but because it could be used to give rides to children and, therefore, pay for itself. The Asian elephants were relatively easy to acquire and one was soon on its way from Calcutta. Within weeks of its arrival in March 1883 it was already giving rides to children. Some months later the Zoo received another elephant, this time from the King of Siam in exchange for Australian fauna; the demand for rides was not sufficient to keep it busy and it was eventually sold to the Adelaide zoo.²

Two other animals which le Souef and the Council wanted for the Melbourne Zoo were the tapir and the rhinoceros. Both of these animals were native to the Malay peninsula and promised to be sufficiently unusual to draw in the crowds.³ They first went in pursuit of these animals in 1884 when Dudley le Souef was sent to Singapore in the hope of securing one of each. He spent an eventful month there, not least because of the hazards of staying in a cheap hotel to save on costs.⁴ However, after several trips to Asia on behalf of the Zoo, Dudley was an experienced traveller and was not distracted from the purpose of his visit. He bought two tapirs and other animals which he sent to Melbourne unaccompanied. No rhinoceros were available for sale at the time and he spent the next month waiting in Singapore for news of one. He amused himself by collecting, killing and preserving snakes for the small museum in the Melbourne Zoo.⁵ Then, just when he was about to give up hope, he received a message that

---

1. St Leon, *The Circus in Australia*, p.vi-viii.
2. ZASV minutes, 23 January 1882, 5 March 1883, 25 June 1883, 26 November 1883. Adelaide Zoo was founded in 1878.
3. The Sumatran rhinoceros is a two horned animal and smaller than African species; the Malayan tapir, from the same area, is related to the rhinoceros family.
4. On one occasion he was driven out of his hotel room by cockroaches and, a few nights later, the guest in the neighbouring room committed suicide (Dudley le Souef's diary in the family archive, entries for the 14 August, 1884 and the 15 August, 1884).
5. The Zoo's museum was founded in 1875, with a collection of birds belonging to F.R. Godfrey forming the nucleus of the collection; it was arranged that Dudley would learn taxidermy and take responsibility for developing the museum (ZASV minutes, 28 September 1875).

a rhinoceros was available in Klang. After yet another delay, during which time he made copious notes in his diary about the local bird calls, he finally succeeded in buying one of the elusive animals. To his dismay the rhinoceros became ill during an unscheduled transhipment in Sydney and died before it reached Melbourne; the shipping company agreed to pay some compensation. To compound his misery, the tapirs which he had sent back unaccompanied also died, one on the voyage and the other shortly after its arrival. The trip had taken three months, had cost £400 but, although the most important animals died, other interesting animals, including a black panther, a leopard, a tiger and some orang outan, survived.[1]

Two years later, Dudley le Souef successfully brought an American tapir back with him from Europe. The rhinoceros now became the major target for the Zoo's administrators. After several unsuccessful attempts, the Zoo's agent in Calcutta, a Mr Saynal, purchased one on their behalf despite strong competition from English buyers. He sent it to Melbourne on the SS Bancoora, a ship which had carried animals for the Acclimatisation Society and the Zoo for many years; but this time the Bancoora was shipwrecked on the way back to Australia. The rhinoceros somehow survived and, a week later, was removed from the ship and was sent on to Melbourne. It proved to be an enormous attraction; despite unfavourable weather the attendances at the Zoo doubled. Sadly, it had sustained injuries during the shipwreck and died some weeks later. Subsequent offers of rhinoceros proved to be beyond the price range of the Zoo and no further attempts were made to acquire one in the nineteenth century.[2]

The Melbourne Zoo's brief encounter with the rhinoceros was a good example of the value of the 'starring' system, a system which had been developed in the London Zoo in the 1850s. In order to increase flagging attendance figures, the Zoo's Secretary, David Mitchell, decided to purchase a new and unusual animal for the collection and publicise it widely. A hippopotamus, which arrived in 1850, was the first 'star'; an elephant and her calf followed, as did an iguana, a chimpanzee and an anteater.[3] In Melbourne these developments were followed with interest. The Council members were greatly impressed with reports of the London Zoo's annual income soaring from £3000 to £18,000 in the months af-

---

1. ZASV minutes, 1 December 1884.
2. ZASV minutes, 7 June 1886, 2 March 1891, 3 August 1891.
3. Altick, *The shows of London*, pp.318-319.

ter the hippopotamus arrived.[1] They placed their hope of emulating that success in acquiring a rhinoceros. Later on, the giraffe and the hippopotamus were included on the list of desirable exotic animals, but both of these were quite beyond the Zoo's price range in the nineteenth century.

The emphasis on 'stars' such as the rhinoceros, giraffe and hippopotamus reflected a popular fascination with the exotic: it was the strangeness, the otherness of these animals which made them so attractive. And, as Mullan and Marvin argue in their study of zoo culture, the attraction of the exotic was perhaps reinforced by its association with childhood; after all it was in childhood stories and folk-tales that most Australians would have been introduced to the strange animals.[2]

But there were other, more subtle, characteristics which increased the exhibition value of particular wild animals in captivity. There was generally a preference for animals which engaged in a high level of activity in their enclosures or which interacted with the public or with themselves. Animals which displayed ferocity or were known to be dangerous, and which were most easily anthropomorphized, also attracted attention.[3] This helps to explain why, in the Melbourne Zoo, monkeys were one of the most popular exhibits; they displayed several of these characteristics, being highly active, easily anthropomorphized and responsive to the public. They were frequently singled out for attention by visiting journalists. 'Liber', writing for the *Age* (23 January 1875), found it 'as amusing to watch the visitors around the monkey house as it is to observe the inmates themselves. The children laugh at their antics, and recognise in them fellow beings. Their elders look amused and smile...while others have a serious look, as though they were admitting to themselves that, perhaps after all, Darwin is right.' Another writer described how three monkeys, chained to a pole, 'were constantly tantalised by handfuls of nuts and biscuits thrown just out of reach,' and how he waited for some time 'to look at the frolics of these droll creatures, with their strange mimicry of us and wondered which of our friends they most resembled' (*Daily Telegraph*, 14 January 1882). Even the Zoo guide book noted

---

1. F.R. Godfrey speaking at the Annual General Meeting, 19 February 1883, in the *Argus*, 27 February, 1883.
2. Mullan & Marvin, *Zoo culture*, p. 3.
3. Mullan & Marvin, *Zoo culture*, p.74.

that the baboons and other monkeys caused 'many a hearty laugh to the old as well as to the young.'[1]

In front of the monkey cages the behaviour of Melburnians was no different, it appears, from that of their European counterparts. In accounting for such antics, several historians have drawn an illuminating parallel with the conduct of visitors to an eighteenth century madhouse. In the London Bedlam, for example, the inmates were considered to be more animal than human; their unrestrained wildness compounded by the conditions in which they lived made them a source of amusement and entertainment.[2] Consequently, Bedlam featured on the tourist's list of the sights of London along with the Menagerie in the Tower. Until the practice was stopped in 1770[3] visitors paid an entrance fee to wander amongst the cages of inmates, some of whom were chained. In describing such a visit, one writer recalled an event in which an apparently docile inmate chewed large quantities of cheese and bread until 'at last he counterfeits a sneeze, and shot such a mouthful of bread and cheese amongst us...which made us retreat.'[4] The danger of standing too close to monkeys in a zoo was a variation on this theme. But there were other parallels too; dull inmates in Bedlam, for example, were goaded into providing more entertaining behaviour by being prodded with walking sticks, or with ridicule, gestures and imitations.[5] Madness, Foucault points out in his study on the subject from the middle ages to the eighteenth century, became a living spectacle.[6]

The primates were not the only animals in zoos which were encouraged to provide more entertainment. Actions designed to provoke the lion to roar and the bear to growl were part of the zoo visitors' routine. Madness was also a thing to be heard.[7] Consequently, the lions, tigers, panthers and other big cats were popular exhibits in the Melbourne Zoo, eclipsing all of the other animals especially during feeding time, when the snarling, growling and other fearsome noises were at their most exciting. The carnivores' feeding time was a highlight of a visit to a zoo; 'Turning away [from the lion cages],' wrote a reporter from

---

1. A.A.C. le Souef, 'Guide to the Zoological and Acclimatisation Society's Gardens' in *Proceedings of the Zoological and Acclimatisation Society of Victoria*, Vol. IV, 1875, p.42.
2. Mullan & Marvin, *Zoo culture*, pp.34-36.
3. Altick, *The shows of London*, p.45.
4. Mullan & Marvin, *Zoo culture*, pp.35-36.
5. Altick, *The shows of London*, p.45.
6. Michel Foucault, *Madness and punishment*, cited in Mullan & Marvin, p.36.
7. Elaine Showalter, 'Victorian women and insanity', cited in Mullan & Marvin, p.36.

the *Daily Telegraph* in 1884, 'we hear behind us a hum of voices and tramping of feet...Here coming down the wide paths towards us some two hundred people. Men with chimney pot hats, and men with roguish-looking deer-stalkers...on one side of their heads, men leading a boy and a girl by either hand, men carrying infants; and women and girls...dressed in all the shades of all colours...On they come, laughing and talking loudly, but still as if on some fixed purpose bent, and we can discover no cause, until the crowd reaches us, parts, and from the middle there emerges a man, pushing before him a hand cart, on which are large joints of beef. Now the truth dawns...the time has arrived for feeding the wild beasts.'[1] The carnivores were not fed on Sundays in front of the non-fee paying audience, a system which probably had less to do with the animals' diet, than with an awareness on the part of the Zoo's administration of the appeal of this exhibition (*Daily Telegraph*, 14 January 1882). 'Liber', speculating on the attraction of carnivores and other deadly animals in the Melbourne Zoo, wondered how many people visiting this and similar exhibitions did so in the hope of seeing a man 'torn to pieces'; 'exhibitions of a dangerous kind,' he observed, 'are far more attractive than those where a broken neck is not a likely contingency' (*Age*, 23 January 1875).

The reptiles generated almost as much excitement for much the same reason, their potential to kill the visitor. In this instance, the Zoo's administrators were more concerned with displaying the reptiles, particularly the Australian snakes, in such a way as to inform about the venomous characteristics of each one (*Argus*, 24 September 1881). But the majority of the public, in search of thrills, had other priorities. One writer noted during a visit to the gardens: 'Reptiles appear to ever exercise a fearful fascination over the popular mind and there is always a large crowd to see the snakes' (*Argus*, 2 December 1876).

At the other end of the spectrum of exhibition value were the birds and fishes.[2] However, even in these lowly creatures, public interest could, it was thought, be stimulated by providing an attractive setting. Fish were put on display in the London Zoo when the first public aquarium was installed in 1853. The public responded well and, in the following decades, many of the Conti-

---

1. *Daily Telegraph*, 2 January 1884: this practice has been stopped in many zoos around the world in recent years because it is felt that the snarling and growling behaviour displayed by the animal is a poor representation of their true behaviour and panders to popular misconceptions (Mullan & Marvin, p.6).
2. Mullan & Marvin, *Zoo culture*, pp.73-74.

nental zoos built their own, sometimes highly elaborate, aquaria with plants, rocks and grottos.[1] Several members of the Council thought that one would be an attractive addition to the Melbourne Zoo and, although the suggestion was raised on several occasions, decisions on the matter were postponed repeatedly. No one thought of combining aspects of the continuing fish acclimatisation activities with the provision of an aquarium.

On the other hand, the birds in the acclimatisation programme appear to have suffered from attempts to make them more accessible to the public. The pond where the waterfowl congregated, an area which le Souef regarded as having 'a most uninviting appearance',[2] was cleared out and the thick scrub cut back in order to provide the public with a better view of the pond.[3] A few years later le Souef and his staff divided the ponds into sections in order to control the waterfowl more effectively.[4] The conditions suitable for most successful breeding for acclimatisation purposes were thus eroded as the exhibition of animals took priority.

One unusual suggestion for an exhibition made during this period was the creation of an insect house. F.R. Godfrey, a senior member of Council, first recommended that it might be an attractive addition to the collection; he had recently visited the insect house in the London Zoo and had been very impressed by the responses of the large crowds to the tiny creatures.[5] The London Zoo insect house comprised both living and dead insects held in cases; the living insects included the silk-moth and the green-fly.[6] Some years later Godfrey again raised the topic: in his opinion 'the time had arrived for establishing an insect house in the gardens as it would be of great practical utility and also a source of interest to visitors.' It was suggested that an insect house could be created with little expenditure in Melbourne Zoo but the idea did not receive much support from other council members who believed that such an undertaking was not an appropriate exhibition in a zoo.[7]

---

1. Hancocks, *Animals and architecture*, pp.149-150.
2. ZASV minutes, 17 May 1872.
3. 'Report of the Council' in *Proceedings of the Zoological and Acclimatisation Society of Victoria*, Vol. II, 1873, p.7.
4. ZASV minutes, 14 June 1875.
5. 'Report of the Council', in the *Argus*, 25 February 1883.
6. Scherrin, *The Zoological Society of London*, p.176.
7. F.R. Godfrey speaking at the Annual General Meeting, 17 February 1890 in the ZASV minutes 17 February 1890.

The International Zoo

Although the zoological collection in the Melbourne Zoo was limited throughout this period by lack of funds, the animals which were exhibited clearly indicate that le Souef and the Council were aware of what interested the public. By 1887 the Zoo boasted a collection of over four hundred animals and more birds. The animals included over sixty carnivores, amongst which were a Siamese cat and a Persian cat, there were also rodents, rabbits, ungulates as well as five thylacines, seven varieties of kangaroo, two platypus and other native animals. The public were responding well and over one third of its income, or about £2,500, was collected at the gate or by subscription.[1] Undoubtedly the trips to European zoos had provided useful information on public taste in animals on exhibition.

The increasing number of imported animals made little difference to the nature of the enclosures the Zoo had already established. The herbivorous animals remained in the paddocks and the carnivores and primates stayed in cages of varying quality, depending on how much money was available for improvements at any given time. Some of the cages had wooden floors, others had asphalt floors; some were incorporated into solid structures which were constructed when money was available, and many of the early ones stood in rows protected only by trees and shrubs.[2] The large North American bears were kept in a traditional bear-pit modelled on the famous sixteenth century bear pit of Berne. The Melbourne bear-pit was 9 feet deep, made of brick and cement, and featuring a 16 foot pole in the centre (*Argus*, 25 December 1875). Shortly after it was completed le Souef admitted that he had made a mistake 'by building one of the old fashioned bear pits which are now, I understand, being altered in the European gardens, and which I find are unsuited to this climate.'[3] It was replaced some years later by a bear house.[4]

The enclosures which caught the attention of le Souef on his trip to Europe in 1880 were the modest animal houses in the Cologne and Rotterdam zoos. He considered them worth noting in his report to the Council because they were 'pretty' and, more importantly, were 'within the compass of what we might rea-

---

1. 'Report of the Council', in *Twenty-fourth annual report of the Zoological and Acclimatisation Society of Victoria*, p.9.
2. For example, in 1873 the two lions were moved into a cage measuring 20 x 8 feet with a small wooden shed attached for an unspecified purpose; this was 'double' the size of the cage they had been living in for a year (ZASV minutes, 4 October 1873).
3. ZASV minutes, 6 October 1876.
4. ZASV minutes, 30 October 1882.

sonably hope our means would be.'[1] What he ignored for the purposes of his report were the large and expensive buildings which had been constructed in some of the wealthier zoos, such as Amsterdam, Antwerp and Cologne. These buildings were not designed with the functions and activities of the animals in mind but, as Hancocks suggests, 'to try and create a mood which seemed sympathetic to [the animal's] legendary history or its country of origin.'[2] One of the best known houses was the elephant house in the Berlin Zoo which was built in the form of a Hindu temple; when Dudley le Souef visited the Berlin Zoo in 1898 he was very impressed by both the zoological collection and the buildings, but most particularly by the elephant house which, he noticed, attracted almost as much attention as the animals themselves.[3]

Although the Melbourne Zoo could not afford to build grand animal houses on this scale, several miniature versions of such houses were erected for small animals. The guinea-pigs lived in a model Swiss chalet with a balcony and curtained windows; next door was a 'formidable Norman castle,' called 'rat castle,' with battlemented towers, a banner and ramparts which housed a dozen white rats; and a log-cabin next to that, surrounded by a flower garden, was for a family of rabbits.'[4] These proved tremendously popular with the public, adults and children alike, and, while they were certainly cheaper than some of the magnificent buildings of European gardens, the intentions of the architectural design were similar.

Zoo design is considered indicative of how zoo architects and administrators regard the proper relationship between man and beast.[5] Current scientific knowledge about animals is reflected in the fashions of exhibition design; as this knowledge changes, the architecture also changes [6]. Reflecting these interests, zoos in the nineteenth century were laid out on a systematic basis with the animals clustered together in taxonomic groups[7] and their enclosures designed to present an unimpeded view of the exhibit.[8] Even those enclosures contained

---

1. Albert le Souef's report to the Council on his return from Europe, 30 August 1880.
2. Hancocks, *Animals and architecture*, p.106.
3. Dudley le Souef's report to the Council on his return from Europe, ZASV minutes, 16 January 1899; the Berlin elephant house was destroyed in 1941 (Hancocks, pp.110-111).
4. 'Report to the Council', ZASV minutes, 20 February 1882.
5. Mullan & Marvin, *Zoo culture*, p. 53.
6. Paul Boussiac, *Circus and culture*, p.110.
7. Hancocks, p.129.
8. Boussiac, p.111.

within large elaborate structures were designed on the same principles; the major difference in conditions for the animals was that they were protected from climatic extremes.

While the intentions and interests of le Souef and the Melbourne Zoo Council were similar to those of other zoo administrators in the nineteenth century, they were not always in accordance with the convictions of the visiting public. There was by no means unanimous acceptance of the practice of holding wild animals in small enclosures. As early as 1875, *Liber*, writing in *The Age* about a visit to the zoo, wondered about how a monkey 'got into his present unfortunate position', and whether the kangaroos, as they 'disport themselves in the limited space allotted to them' were 'thinking of the wild freedom of their youth.' He speculated on the condition of the lions' eyesight and feet, as 'their constant walking backwards and forwards before those iron bars must in time have a bad effect upon their sight, and the hard unyielding boards must be ill-suited to feet intended by nature to walk over the springy sod' (*Age*, 23 January 1875). A year later, 'The Vagabond' quoted an article from the *Edinburgh Review* of 1855, in which the writer asked of lions and tigers: 'Why do we coop these noble animals in such nutshells of cages?' They should, he suggested, be held in 'half an acre of enclosed ground strewed with sand...and with 20ft of artificial rock.' 'The Vagabond' pointed out that such an idea could be used in the Melbourne Zoo because of the amount of space available for the animals. He added that the Zoo Council could gain financially by allowing the animals into these enlarged enclosures at designated times so that the public could watch them 'bound' (*Argus*, 2 December 1876).

The criticisms continued as the international collection grew. A social writer, who was more interested in the dress sense of the female visitors to the Zoo than in the animals, made a brief and sympathetic reference to the newly acquired Bengal tiger: 'his temper was anything but good, for he watched the people gazing at him with a savage look, as though their curious stares added the last touch of bitterness to his ignoble bondage.' As for the fashions, she considered that there were 'no dresses worth describing, as most wore dust-coats or simple print' (*Herald*, 23 January 1882). Even the parliamentarians, when shown the large carnivora during one of their annual visits, expressed a desire that the Director should give the 'poor beasts' a 'run' in the grounds (*Herald*, 28 October 1885). In 1890 a journalist remarked that 'all lovers of animals pity the confined position of the large carnivora, whose ceaseless walk and restless eyes tell of un-

The International Zoo

curbed energy and desire for freedom' (*Age*, 22 February 1890). Yet there was an ambivalence in the attitude of the critics as these comments were usually interpolated into lengthy descriptions about the Melbourne Zoo which were otherwise fulsome in their praise.

The Zoo's administrators did not comment on the criticisms in the Melbourne newspapers. There was a growing sensibility towards animals in the nineteenth century, evident in Australia with the foundation of the Society for the Prevention of Cruelty to Animals (SPCA). The Victorian SPCA, founded in 1870 and modelled on the British SPCA,[1] was selective in its work. It spent much of its time trying to curb the abuse of horses by cab drivers, and it mounted education programs in order to discourage 'deliberate wickedness, thoughtlessness, and ignorance.' Yet at the same time it supported other blood sports: '[men] shot and hunted and fished, not for the pleasure of killing, but for the pleasurable excitement which accompanied those pursuits; and so long as men hunted in a sportsmanlike way, giving the creature a chance for its life, [the chairman, George Verdon] could not see objection to it.'[2]

The extent to which the slowly developing sensibility about animals in society influenced the journalists' comments about the animal enclosures is open to conjecture. The tone of many of the journalists' comments suggests that much of the sentiment may have come from an anthropomorphic response to seeing animals held in cages. Not all animals received this attention; and it was the lions, 'kings of the jungle', which were most often cited in this regard. Certainly the Melbourne writers did not suggest returning the animals to the wild, nor did they appear to question the validity of zoos generally. Little was then known about animal behaviour in natural habitats in the nineteenth century.[3] Rather, their ambiguous attitude may be explained by their presumption of the educational role of the zoo, which would provide sufficient justification to keep wild animals in captivity.

Formal education in the Melbourne Zoo in the nineteenth century, in so far as it existed, consisted of labels containing basic information, including the

---

1. Founded in 1824; the concept spread throughout the western world during the nineteenth century.
2. Victorian Society for the Prevention of Cruelty to Animals, Annual Report, 1874, p.6-7.
3. Ethology gained international acceptance when Konrad Lorenz and Niko Tinbergen, both zoologists, received a Nobel Prize for medicine in 1973; they studied animals in their natural environment in order to overcome anthropomorphic analysis of animal behaviour (*Oxford companion to animal behaviour* (New York, 1987), p.153).

popular and Latin names of an animal and its approximate area of origin. The arrangement of animals in taxonomic groups and in cages which permitted a full view of the exhibit was considered adequate provision for comparative analysis of varieties within species. If the aim of zoology was 'to furnish every possible link in the procession of organised life', as *Vagabond* described it, then a zoo was clearly the place to study it (*Argus*, 2 December 1876). The Melbourne Zoo guide book produced in 1875 contained a combination of information including notes about the source and physical description of the animal, as well as anthropomorphic descriptions, such as 'the hideous Tasmanian devils.' It also commented on comments public reactions, as for example, a reference to the monkey houses, pointing out that they 'attract a great deal of attention and cause much amusement among juvenile visitors.'[1] Newspapers occasionally wrote a similar combination of information on the arrival of new stock. In 1883, for example, the *Argus* and the *Daily Telegraph* both recorded the results of Dudley le Souef's recent trip to the United States in some detail; after noting the Latin name and the nocturnal habits of the racoon, the *Argus* described it as 'a most amusing animal,' and hoped that 'no larrikin may think proper to put out the eye of this new racoon with a stone, which was the treatment accorded to the last' (*Argus*, 1 October 1883). The *Daily Telegraph* drew attention to the civet cat and its 'really nauseous smell'; it went on to reassure its female audience that when the civet was removed from the pouch of the cat and treated properly, it produced the pleasant fragrance with which they would have been familiar (*Daily Telegraph*, 6 October 1883).

The presentation of information in the Melbourne Zoo was similar to that in most nineteenth century zoos. Some zoos, most notably the London Zoo and the *Jardin des Plantes*, had more sophisticated education programs which incorporated lectures and journals and which were aimed at specialised audiences. On one collection trip, le Souef observed that the *Jardin des Plantes*, which was a combined zoological and botanical garden, had a staff of 17 professors and was operated 'with the object of instruction rather than amusement.' He observed that 'when an animal or bird died...its chief use began, as it then furnished a subject for lecture or specimen or both'[2] Much of the education, however, appeared to go on in large lecture halls and was directed towards students of zool-

---

1. A.A.C. le Souef, 'Guide to the Zoological and Acclimatisation Society's Gardens' in *Proceedings of the Zoological and Acclimatisation Society of Victoria*, Vol. IV, 1875, pp.37-38.
2. ZASV minutes, 30 August 1880.

ogy. Entertainment was not a priority of the administrators of either the London Zoo or the *Jardin des Plantes*, yet they were regarded as the finest zoos by commentators in Melbourne and represented the standards to which their local Zoo was encouraged to aspire.[1]

Despite the inadequacy of the Melbourne Zoo's educational program in modern terms, the importance and usefulness of education in the Zoo were frequently mentioned in the nineteenth century. Le Souef declared his intention of making the Zoo gardens 'of value to the colony from an educational and scientific point of views.' George Bruce, when he was president of the ZASV, considered that the Zoo 'ranked as a national educational institute' (*Daily Telegraph*, 28 November 1888). Members of Parliament agreed: Graham Berry regarded the Zoo as 'a most interesting mode of amusement and instruction' (*Herald*, 28 October 1885); another MP, Mr Levien, referred to the gardens as 'a public school' (*Argus* , 25 October 1889). The *Argus* also discussed the educational value of the Zoo: 'In a collection of this nature, instruction and amusement are so happily combined that one is enabled to imbibe valuable doses of natural history with an ease which could not otherwise be provided' (*Argus*, 27 March 1883). Without the Zoo, it noted some years later, 'literature must lose some of its intelligibility to [young people]' (*Argus*, 14 October 1893). One writer went as far as to suggest that a zoo 'might be properly attached to the Education department, and...a good deal of money now wasted in over educating or vainly educating youths and maidens for very ordinary work in life would be better spent in support of this great instructive and entertaining institution' (*Argus*, 15 April 1889). The *Herald* summarised the role of a zoo succinctly: 'There be some people, doubtless, who are on principle opposed to detaining any animals in captivity, but the large majority agree in thinking that collections of zoology are a valuable and important adjunct to practical education' (*Herald*, 28 November 1888).

The educational content of the Zoo was, quite simply, the presence of the live animals. Observations recorded by contemporary commentators hint at the way visitors responded to the Zoo's exhibits on an intellectual level. A child's interest was summarised by *Sketcher* : 'Witness the delight with which a group of intelligent country children, well instructed, view the Society's zoological collection for the first time. Knowing most of the animals from pictures, and from the

---

1. For example *Herald*, 28 November 1888.

descriptions they have read in natural history books, they take the liveliest pleasure in seeing them in the flesh, and recognising them. The experience gives sanction and authority to ...book learning, and will certainly make them more attentive and trustful readers for the future' (*Australasian*, 12 December 1874).

Adults, according to various sources, could ponder Darwin's theory of evolution while visiting the Zoo. Although Darwin's central thesis was not generally accepted in Australia until the 1890s, it was the subject of debate and speculation.[1] 'The Vagabond', for example, noted that on a visit to the Zoo, 'the origin of the species may be studied, that progression of human life which ends in man and assumes no higher form' (*Argus*, 21 July 1884). And, an *Age* reporter suggested that 'to the Darwinian a zoological garden is a history of his ancestors, replete with thrilling genealogical interest' (*Age*, 13 October 1893). Some zoos in Europe exploited this interest in evolution. Le Souef described a display he saw at the French Acclimatisation Gardens in the Bois du Boulogne: 'a grand panorama of the pre-Adamite world in which all the old and now extinct animals of the earth are depicted...the scenery is also of the character which scientists tell us existed in those days.' The closest the Melbourne Zoo got to that sort of exhibition was a large model of an extinct Australian marsupial made by the confirmed anti-Darwinian, Frederick McCoy. The model, which was placed in an enclosure beside the live kangaroos, was made as part of the Zoo's contribution to the 1888 centenary celebrations; most certainly, it was not there to demonstrate the theory of evolution.[2] The *Argus* called for a more complete display: 'it would be wonderfully instructive and amusing to have a collection of our prehistoric monsters reproduced' (*Argus*, 15 April 1889); the Zoo's administrators, however, did not respond.

The one area where the Melbourne Zoo administration appears to have taken an active approach to education was that of the native fauna. Although le Souef and his staff were not prepared to become involved in the study of the embryology of marsupials, they used opportunities to educate the public on the range of native fauna. Indicating the venomous qualities of the reptiles and separating the different varieties of kangaroo into individual enclosures for easy identification stemmed from a genuine desire to increase local knowledge of indige-

---

1. Moyal, *Scientists in nineteenth-century Australia*, pp.188-189.
2. ZASV minutes, 30 July 1888.

nous fauna.[1] The attempts at education obviously failed to inspire popular interest in the native fauna held in the zoo because the imported animals continued to command the local people's attention. An *Age* reporter even implied that it was more important to have a good collection of foreign animals in Melbourne because indigenous animals were so dull (*Age*, 13 October 1893).

Like other nineteenth century institutions of collection and exhibition, such as museums and galleries, zoos frequently justified their existence in terms of education. Yet in comparison with museums and galleries, their cultural status was low. One explanation for this, according to an argument expounded by Mullan and Marvin, is that animals cannot attain the level of cultural importance which is accorded to man-made objects in, for example, a museum. Exhibits in museums require interpretation, they have meaning in themselves, and have symbolic importance in the study of the development of a civilization. Even stuffed animals in a natural history museum and man-made objects representing animals have a higher cultural status than live animals. Animals have no meaning in themselves; they simply exist.[2] A greater concentration on education at the expense of exhibition value may have rectified this imbalance, but le Souef and the Council could not afford to do that, as education in itself was unlikely to attract a large paying audience.

The provision of entertainment in the Zoo extended beyond the animals. The additional amusements and catering facilities in European zoos greatly impressed le Souef on his trip there in 1880.[3] Rotterdam Zoo's restaurant, he said, was 'palatial in proportion and elegant in design' and, during the summer months, concerts were organised and attended by thousands of people. Additional attractions were laid on by most zoo administrators in the nineteenth century with the intention of creating a resort to which visitors would return repeatedly. Zoos pursued these activities to different degrees; at one end of the scale the older institutions, such as the London Zoo, had simple facilities and entertainments, and at the other end the German commercial zoos boasted extensive facilities ranging from bars and restaurants to game and party rooms.[4] Enviously, le Souef reported on the incomes generated by some of these grander facilities,

---

1. A.A.C. le Souef, 'Guide to the Zoological and Acclimatisation Society's Gardens' in *Proceedings of the Zoological and Acclimatisation Society of Victoria*, Vol. IV, 1875, pp.43-44; ZASV minutes, 6 December 1886.
2. Mullan & Marvin, *Zoo culture*, p.116ff.
3. ZASV minutes, 30 August 1880.
4. Mullan & Marvin, *Zoo culture*, p.111.

but he did not have the resources to pursue such luxuries for the Melbourne Zoo.

The inclusion of facilities and entertainment in zoos highlighted their connection with the pleasure gardens of previous centuries, a connection which was traced by Mumford to the influence of the Baroque palace. He suggested that pleasure gardens, such as the Vauxhall Gardens in London, stemmed from 'attempts to supply the more lascivious pleasures of the court to the commonalty at a reasonable price per head.'[1] In the 1830s the London Zoo and the Surrey Zoological Gardens, a commercial enterprise founded in 1831, provided direct competition to Vauxhall Gardens.[2] The design of the Surrey Zoological Gardens was similar to that of traditional pleasure gardens with promenades, flower beds, statuary, fountains, grottoes and cascades. Its administration was aware that animals alone would not be sufficiently interesting to encourage visitors to return frequently and, being entirely reliant on the entrance charge, they organised additional attractions to entice the crowds.[3] The colonial parallel was the Cremorne Gardens in Richmond. Modelled on the Vauxhall Gardens in London, they opened in 1853 and expanded under Coppin's management in 1856 to include a dance-floor, a menagerie and a bandstand; he also organised various spectacular shows usually accompanied by fireworks. Coppin sold the Cremorne Gardens in 1862, long before they could be seen as a challenge to the entertainments in Royal Park.[4]

During the nineteenth century the Melbourne Zoo's additional facilities to amuse and entertain visitors did not extend beyond relatively simple catering facilities, animal rides, a band and an ethnological exhibition. In the 1860s and 1870s provision for the comfort of visitors was rudimentary, with a refreshment tent and some benches. But when the admission fee became a reality in 1881, improved refreshment facilities became one of the aims of the administration. Le Souef had observed that the European zoos attracted the 'best families' and 'respectable people' and he was anxious to draw in the same kind of clientele to

---

1. Mumford, *The city in history*, p.379.
2. Altick, *The shows of London*, p.323.
3. Various spectacular shows including Vesuvius erupting, a modelled view of Rome, the siege of Gibraltar, the siege of Badajoz preceded by a fireworks display, and Napoleon crossing the Alps were amongst the subjects of the elaborate displays which attracted large numbers of visitors to the Surrey Zoological Gardens between 1837 and 1850 (Altick, pp.323-327).
4. Serle, *The golden age*, p.364; Grant & Serle, *The Melbourne Scene*, pp.107-108; the buyer converted the Cremorne Gardens into a private lunatic asylum (*ADB*, v.III, p.461).

the Melbourne Zoo. Thus, in 1885, a refreshment house 'of superior class' was built which provided lunch. After some discussion, the Council decided against applying for a liquor licence for fear that it might encourage excessive drinking.[1] Instead, it opened a milk bar by converting the old refreshment house and tethering a cow nearby to provide the fresh milk.[2]

Bands made occasional appearances in the Zoo. The most sustained effort was a series of concerts organised in 1885 by the 'Coffee Garden Society', which attempted to create an atmosphere similar to that in European, particularly German, gardens. The concerts continued spasmodically over the next three years but were eventually discontinued because they were not profitable.[3] A year later, the *Argus* bemoaned the absence of a band, stating that 'there should be a band at the Zoo on every day. We want state-paid bands for the public amusement' (*Argus*, 15 April 1889). Although le Souef brought several other bands into the Gardens on occasion, none attracted the crowds which he had anticipated.

Elephant and camel rides were most popular entertainments amongst the junior visitors to the Zoo. The Melbourne elephant, referred to by the *Daily Telegraph* (2 January 1884) as 'our Victorian Jumbo', began carrying children in 1883 on a howdah designed to hold six children at a time. Within a year of its arrival, the animal, named 'Ranee' by the Zoo's staff, had carried over fourteen thousand children on its back.[4] The camel, which could carry four children at a time, did not seem to attract as much attention as the elephant. Perhaps the apparent instability of the camel, particularly when it knelt down, frightened the young visitors. The rides were financially successful, producing an average annual income of £140 even during the lean years towards the end of the century. Undoubtedly, the elephant and the camel contributed significantly to a claim made by the *Evening Herald* (5 March 1883) that 'a visit to the Zoo on any day is estimated by the majority of children as the greatest treat a parent can afford them.'

---

1. ZASV minutes, 19 April 1886.
2. 'Report of the Council', in the *Argus*, 22 February 1887.
3. ZASV minutes, 2 November 1885, 5 December 1892.
4. 'Report of the Council' in the *Argus*, 5 February 1884. Jumbo, London Zoo's famous African elephant, had been used to give rides to children between 1865 and 1882 (Blunt, *The ark in the park*, 178ff).

The International Zoo

One variety of side show which was becoming popular in Europe was the animal act. Dudley le Souef described a performance he watched in the Berlin Zoo in 1898 involving animals trained by Carl Hagenbeck: a keeper strolled amongst a group of lions, tigers, polar bears, hyænas and other potentially dangerous animals, playing with them and feeding them for the entertainment of visitors.[1] Although Melbourne did not introduce this kind of act until the twentieth century, one of its staff, Mr Jensen, had a short routine of his own which he performed for select visitors, whereby he 'kissed the male lion and put his head into its mouth' (*Argus*, 25 December 1875). This sort of interplay between the keeper and the animals for the public's entertainment was not uncommon in the nineteenth century. But, unlike the baroque dramatization which surrounded circus acts, animal acts in zoos were low key interactions between animal and its keeper.[2]

The most ambitious side show put on in the Melbourne Zoo in the nineteenth century was the ethnographic exhibition of Aboriginal life. In 1882 le Souef erected 'a native encampment of the olden times such as the blacks used to erect 40 years ago.'[3] It aroused great interest but, although he planned to expand it, he could not strip enough fresh bark and the exhibition gradually disintegrated. It was revived again in 1887 in preparation for the anticipated influx of visitors coming to Victoria for the centenary celebrations and for scientists arriving for the first Australian Association for the Advancement of Science congress scheduled for 1888. The exhibition represented a corroboree of the 1840s in miniature, with bark huts, native weapons and forty or fifty wooden figures two feet high. It was all designed by le Souef, who still had vivid memories of corroborees he had seen as a youth on the Aboriginal protectorate station in Goulburn. As an additional attraction, he persuaded some Aborigines from the Coranderrk reserve near Healesville to set up an encampment and throw boomerangs in the gardens for a few weeks.[4]

Ethnographic exhibitions based entirely on real people rather than models were not unusual in zoos in the nineteenth and early twentieth centuries. Carl Hagenbeck was one of the first to bring people from foreign countries to Europe

---

1. Dudley le Souef's report to the Council on his return from Europe, 16 January 1899.
2. Boussiac, *Circus and culture*, p.109-110.
3. ZASV minutes, 9 January 1882.
4. ZASV minutes, 17 April 1882, 16 January 1888, 30 July 1888; A.A.C. le Souef, 'Personal recollections', p.12. A photograph of the encampment in the Zoo suggests that a family from Coranderrk was included in the display.

to be exhibited alongside the animals in zoos when he brought a family of Lapps to Germany with some reindeer in 1874. The Lapps were not treated as freaks, nor did they perform for the audience during their short visit; they simply 'behaved just as though they were in their native land.'[1] Commercially, Hagenbeck's new venture was highly successful and, over the next sixty years, he introduced his European visitors to Eskimos, Somalis, Indians, Patagonians and Hottentots. Other zoos followed his example and in the late nineteenth and early twentieth centuries many of them had their own ethnographic exhibitions.[2] Le Souef's interest in arranging the native encampment and boomerang throwing exhibition appeared to be a cross between financial expediency and a genuine interest in introducing the Aboriginal culture to foreign visitors.[3] Although he judged the display a success[4] few newspapers commented upon it. One writer who did notice it considered it to be 'one of the most striking examples of the progress which Victoria, in the short space of 53 years, has made.'[5] It is unlikely that le Souef intended to portray that particular message.

By the end of the century, the Melbourne Zoo was a well established institution with a national reputation. As a mark of this achievement, Ernest le Souef was invited to be the first director of Perth's new zoo. He had finished studying veterinary medicine, specialising in care of imported animals, and, with his father's permission, he took up the post. Albert's fourth son Sherbourne, who was studying veterinary medicine when Ernest left, took over his older brother's role in the Melbourne Zoo.[6] Dudley, by this stage, had married and settled in a house built for him by the Council in the Zoo. There was no doubt but that he would take over control of the Melbourne Zoo when his father retired.[7] The Council of the Zoological and Acclimatisation Society of Victoria

---

1. Carl Hagenbeck, *Beasts and men*, cited in Mullan & Marvin, *Zoo culture*, p. 87.
2. Mullan & Marvin, *Zoo culture*, p.86 ff; in their lists of different ethnographic groups exhibited in Europe, they do not mention Australian Aborigines.
3. ZASV minutes, 30 July 1888.
4. ZASV minutes, 5 October 1896.
5. ZASV minutes, 15 May 1888.
6. Albert's his third son, John, died some years earlier; Ernest took up the position of director of the Perth Zoo in 1897 (ZASV minutes, 31 May 1897).
7. Dudley took over as director of the Melbourne Zoo in 1902 on the death of his father; he held the position until he died in 1923 as a result of an assault during a robbery; Ernest held the position of director of the Perth Zoo until his retirement in the 1939; Sherbourne went on to become curator of the Moore Park Zoological Gardens in 1903, the precursor to the Taronga park Zoo; he later took up an administrative position in the Taronga Park Zoo.

was still the governing authority of the Zoo and its members appeared to remain enthusiastic about its duties. But Albert le Souef was unquestionably the most influential administrator in the Zoo and he had groomed Dudley to assume the same role.

In twenty years le Souef had transformed the Melbourne Zoo into a well established institution of international standard. He had taken the same steady approach to his work which had been evident in the 1870s. The collection was not as diverse as many of those in Europe and it contained no animals with 'star' quality. But le Souef was always pragmatic about what the ZASV could afford at a given time and, although he pursued the acquisition of a rhinoceros with determination, he never seemed perturbed by the absence of an exciting addition. As a result of well calculated investigation into international trends, all of the features of the best contemporary zoos had been incorporated into the Melbourne Zoo, if only in a most rudimentary fashion. He had created the structure and it was up to future administrations to develop it as their resources allowed.

Conclusion. The Legacy of the Nineteenth Century

At the end of the century the Melbourne Zoo was a stable metropolitan institution. It had been shaped by the events, decisions and ideas of the Zoo's early administrators in response to the public demands and expectations of the previous forty-three years. Civic pride, colonial development, acclimatisation and recreational needs were catered for by a largely voluntary group of men with very limited resources. As public expectations changed, the Zoo changed with them and, by 1900, it was composed of elements that had been introduced during the different phases.

The most constant feature of the Zoo throughout the nineteenth century was the presence of Australian animals. From the first donation in 1858, these had formed the bulk of the zoological collection. They had been kept for use as international exchange currency as well as for the edification and curiosity of the local population. Later on, as the recreational function of the Zoo became more prominent, the Zoo administration, under the guidance of Albert le Souef, continued to promote and develop the Australian collection. The determination to pursue this course was in defiance of international trends, which made no provision for indigenous animals in commercial zoos. Undoubtedly, the decision in Melbourne to hold local fauna was partially based on expediency; the animals were cheap both to acquire to maintain. But, with some encouragement from a section of the community, there was a genuine intention on the part of le Souef and the Council to promote public interest in kangaroos, wallabies, dingos, koalas and other Australian animals. Undeterred by the failure to inspire popular interest, they continued to uphold the policy. Australian fauna have remained an important feature in the Melbourne Zoo.

The carefully landscaped environment was also a result of the periods of acclimatisation and impoverishment in the 1860s and the 1870s. The size and layout of the reserve in Royal Park were initially designed to accommodate herbivorous animals awaiting acclimatisation in Victoria or transport abroad on exchange. As soon as le Souef became involved in the Zoo, he recognized that pleasant gardens would attract the public and began the gradual process of altering the grounds by laying out walks, planting flower beds, trees and shrubs, and forming ponds and lakes. By 1880 the Zoo bore a closer resemblance to a pleasure garden than it did to a farm – and twenty years later, as the trees and plants

The Legacy of the Nineteenth Century

matured, the ruggedness associated with the acclimatisation function had disappeared. In the late twentieth century, the value of the spacious grounds and of the generous size of some of the enclosures emerged as Zoo staff created gardens free of the clutter characteristic of other zoos around the world.

The Melbourne Zoo's administration did not adopt the features of the standard international zoo until the 1880s. Although attempts had been made in 1858 and 1866 to begin a collection of exotic animals, the perceived needs of a growing colony took precedence on each occasion. By 1880 the demand on the Zoo to fulfil a recreational function was evident as Melbourne moved into a period of prosperity. Le Souef and his sons studied the development of zoos abroad and planned the expansion of the collection carefully. Within the limits of their financial resources, they chose particular foreign animals on the basis of their apparent exhibition value. The most popular animals were usually big vertebrates from Africa and Asia with characteristics that appealed to a large, undiscerning audience. The Zoo's administration spent much of its available resources in pursuit of these animals, and its efforts were not always successful. As with most zoos of the period, the exotic collection remained the unchallenged principal feature of the Melbourne Zoo.

But in contrast to many European zoos, the foreign animals' enclosures at the Melbourne Zoo were not a significant feature in the nineteenth century. Le Souef was well aware of the impact that the huge and elaborate buildings in European zoos made on the visitor, but the level of expenditure that was necessary to construct such a building was well beyond the scope of the Melbourne Zoo administration. Some simpler buildings were erected to house the primates and the carnivores in order to provide them with a fairly constant temperature. But, whether an animal's enclosure was in a building or in a row of unheated cages, its living space was designed to contain it securely and to provide the public with an unrestricted view. The most surprising aspect to emerge in researching the Zoo's enclosures in the nineteenth century was that they were not unanimously acceptable to the contemporary Melbourne audience. Mild criticisms of the enclosures were included in a significant percentage of newspaper reports about visits to the Zoo. Other than that, however, there was no sustained public debate on the matter and no evidence of the Zoo's administration responding to the comments. In the twentieth century the enclosures would become closer in style to their European counterparts. Some of the principles of design have changed, for example slightly restricted views have become more common in an attempt to allow the animals the option to vanish from the public gaze; nevertheless, control of the animals in small areas for the benefit of an audience continues to dominate the zoo architects' brief.

All of the activities introduced during the 1880s and 1890s were designed to draw in the paying visitors; hence the emphasis was on providing entertainment value. With the addition of refreshment facilities and entertainments, the Zoo became known as a place where a family could enjoy a pleasant day out at a reasonable cost. However, the Zoo was also known as an educational institution which was fulfilling its role admirably. Although there were no formal educational programmes in evidence, the display of animals from inaccessible regions of the world made it a useful forum. In addition, the large and fairly comprehensive collection of native fauna had an important educational role, even if foreign visitors appreciated it more than the local people. A disappointing aspect of education in the Zoo was that it never became involved in any of the important zoological issues demanding extensive research, which might well have been accommodated in its grounds.

The Melbourne Zoo, as it emerged in 1900, was very mush an Albert le Souef creation. It was his manipulation of the progress of the institution that transformed the grounds of the Acclimatisation Society into a modern zoo. His clever use of resources and intelligent study of European trends, together with his own interest in native fauna, ensured a smooth transition to a modern Australian zoo. Furthermore, he made sure that his style of operation would continue through the next generation of the Zoo's administration by grooming his sons to take over from him. Albert le Souef's contribution was recognised by the Council after his death in 1902 when it recorded that 'under his untiring care the Gardens had become the best known and complete in the Southern Hemisphere.'[1]

---

1. 'Report to the Council', *Zoological and Acclimatisation Society of Victoria annual report*, 1903, p.7.

Appendices

Appendix 1. Objects of the Zoological Society of Victoria
(Argus 1 March 1858)

Firstly, the introduction and improvement of domestic birds and animals, for which exhibitions shall be held periodically within Melbourne, and prizes awarded.

Secondly, for the importation, care, and domestication of mammalia, fishes, birds and reptiles of this and other countries, more particularly those of rare uncommon species.

Thirdly the encouragement and importation of singing-birds, and the endeavour to propagate them in the country.

Fourthly the obtaining a grant of land from the Government for the purposes of the Society.

Fifthly it is contemplated that a portion of the gardens be set apart and arranged for the care and protection of birds, mammalia, etc., which private individuals may import into the colony, in order that by particular care and attention they may become acclimatised, at such terms as may be agreeable to the discretion of the Society.

Appendix 2. Foundation members of the Zoological Society of Victoria
(ZSV minutes)

Present at the first meeting of ZSV
F.M. Selwyn, President, [1857]
Capt Stoney, Hon. treasurer and secretary.

F.G. Moule   R.G. Alleyne
Dr Thomas Black   William le Souef

H.M. Taylor   George Cole
Richard Nash   Thomas Lempriere
W.H. Archer   Edward Wilson
Henry Creswick   Hugh Chambers
Edgar Ray

Present at subsequent meetings of the ZSV

W. Stubbs   William Blandowski
W. Meek   Alderman Walsh
J.B. Payne   George Coppin
Albert Purchas

Senior appointments made in January, 1858

Sir Henry Barkly, Patron   Mr Justice Barry, President
D.S. Campbell, M.L.A., Vice-president

Appendix 3. Original members of the Board of Management
(ZSV minutes, 24 July 1858)

Dr Evans, M.L.A.	D.S. Campbell, M.L.A.
Professor McCoy	F.M. Selwyn
Dr Mueller	W.A. Archer
Hon. Capt Pasley, R.E.	Dr Thomas Black
Dr Thomas Embling	Hugh Chambers

Appendix 4. Objectives of the Acclimatisation Society of Victoria
(*The rules and objects of the Acclimatisation Society of Victoria*, (Melbourne, 1861, p.3)

- The introduction, acclimatisation, and domestication of all innoxious animals, birds, fishes, insects and vegetables, whether useful or ornamental.
- The perfection, propagation, and hybridisation of races newly introduced or already domesticated.
- The spread of indigenous animals, etc., from parts of the colonies where they are already known, to other localities where they are not known.
- The procuring, whether by purchase, gift, or exchange, of animals, etc., from Great Britain, the British colonies and foreign countries.
- The transmission of animals, etc., from the colony to England and foreign parts, in exchange for others sent thence to the Society.
- The holding of periodical meetings, and the publication of reports and transactions, for the purpose of spreading knowledge of acclimatisation, and inquiry into the causes of success or failure.
- The interchange of reports, etc., with kindred associations in other parts of the world, with the view, by correspondence and mutual good offices, of giving the widest possible scope to the project of acclimatisation.
- The conferring of rewards, honorary or intrinsically valuable, upon seafaring men, passengers from distant countries, and others who may render valuable services to the cause of acclimatisation.

Appendices

Appendix 5. Council of the Acclimatisation Society, 1861

(*The rules and objects of the Acclimatisation Society of Victoria*, p.14)

Edward Wilson, President
Ferdinand Mueller, Vice-president

Mr J. Alves
Colonel Anderson
Mr J.P. Bear
Mr S.H. Bindon
Dr Thomas Black
Mr D.S. Campbell
Mr Hugh Chambers

Mr E. Cohen
Dr Thomas Embling
Mr William Lyall
Professor McCoy
Mr Albert Purchas
Mr H.L. Taylor
Mr W. Watts

Appendix 6. Council of the Acclimatisation Society of Victoria, 1870

(*Report of the Council*, in the *Argus*, 11 March 1871)

Dr Thomas Black, President
Dr von Mueller, Vice-president
Professor McCoy, Vice-president
T.J. Sumner, Hon. Treasurer

Hon. A. Michie
Count de Castelnau
F.G. Moule
Hon. Dr Dobson, M.L.C.
H.P. Venables
Dr Joseph Black
George Coppin

John Steavenson
J.B. Were
F.C. Christy
Albert Purchas
George Sprigg
Robert Hammond
Curzon Allport

Albert le Souef, Hon. Secretary

Appendices

Appendix 7. Council of the Zoological and Acclimatisation Society of Victoria, 1886

(*ZASV, 22nd annual report, 1885, p.4*)

Hon. C.J. Jenner, M.L.C., President
Hon. Robert Simson, Vice-president
F.R. Godfrey, Vice-president
Albert Purchas, Hon. Treasurer

George Bruce
John Halfey
F.G. Moule
C.M. Officer, M.L.A.

C. Ryan
Wm. Robertson
Hon. R.D. Reid, M.L.A.
J.C. Tyler

A.A.C. le Souef, Director

Appendix 8. Council of the Zoological and Acclimatisation Society of Victoria, 1900

(*ZASV, 36th annual report, 1900, p. 5*)

Hon. F.S. Grimwade, M.L.C., President
Dr C. Ryan, Vice-president
D.N. McLeod, M.P., Vice-president
T.R. James, Hon. Treasurer

F.R. Godfrey
C.M. Officer
H.R. Hogg

Hon. R.W. Best, M.P.
Major Albert Purchas
G.W. Bruce

A.A.C. le Souef, Director

BIBLIOGRAPHY

OFFICIAL CONTEMPORARY SOURCES

Manuscript sources
Acclimatisation Society of Victoria, minute books, 1861-1872.
Board of Management, Zoological Committee, minute book, 1858-1861.
Zoological and Acclimatisation Society of Victoria, minute books, 1872-1902.
Zoological Society of Victoria, minute book,1857-1858.
Le Souef family archive including diaries of Dudley le Souef's animal collection trips to Asia and Europe in the 1880s and 1890s, and the *Personal recollections* of Albert le Souef.
Archival material relating to the Botanic Gardens, 1857-1861, Public Records Office, series 1189, 744-748.
Chief Secretary's papers, 1857-1900, Public Records Office.

Parliamentary and other official papers
An Act to Provide for the Preservation of Imported Game and during the Breeding Season of Native Game, 24 Vic., No. 161.
'Acclimatization Society [sic], Return to an order to 1st July 1864', *Victoria. Parliamentary Papers* (1864-5), no. C12.
'Annual report of the Government Botanist and Director of the Botanic and Zoologic Garden', 1859-1861, *Victorian Parliamentary papers*.
'Report from the Select Committee of the Legislative Assembly upon The Zoological and Acclimatisation Society of Victoria Incorporation Bill.' Melbourne, 1884.
'Report from the Select Committee of the Legislative Council on the Alpaca', *Victoria. Parliamentary papers* (1855-6), no. D11a, p.1, 4.
'Report from the Select Committee of the Legislative Assembly upon Live Stock Importation', *Victoria, Parliamentary Papers* (1856-57), Vol.2, no.D15a, p.6.
Victorian Hansard, 1857-1900.
The Zoological and Acclimatisation Society Incorporation Act, 48 Vic., No. 794.

OTHER CONTEMPORARY SOURCES

Sources published by the Melbourne Zoo
Acclimatisation Society of Victoria, *Rules and objects*. Melbourne, 1861.
Annual report of the Acclimatisation Society of Victoria, 1862-1867.
Bennett, George, *Acclimatisation: its eminent adaptation to Australia*, a lecture delivered in Sydney and republished by the Acclimatisation Society of Victoria. Melbourne, 1862.

Buckland, Frank, *The Acclimatisation of Animals*, a paper delivered to the Society of Arts, London, republished by the Acclimatisation Society of Victoria. Melbourne, 1861.

On the true principles of breeding developed in letters by Chas H, MacKnight and Dr Henry Madden. Melbourne, 1865.

Annual report of the Zoological and Acclimatisation Society of Victoria, 1885, 1887, 1894-1910.

Proceedings of the Zoological and Acclimatisation Society of Victoria, Vol. I-IV, 1872-1875.

Other sources

Transactions of the Philosophical Institute of Victoria, Vols. I-III, Melbourne, 1857-1859.

Transactions of the Philosophical Society of Victoria. Vol. I. Melbourne, 1855.

Transactions of the Royal Society of Victoria. Vol. V. Melbourne 1860.

Victorian Society for the Prevention of Cruelty to Animals [*proceedings of annual meetings*] 1871-1900.

Newspapers and periodicals

Age (Melbourne).

Argus (Melbourne).

Australasian (Melbourne, 1864-).

Daily Telegraph (Melbourne).

Herald (Melbourne).

Yeoman and Australian Acclimatiser (Melbourne, 1861-1864).

Books and articles

'Garryowen' (Finn, E.), *The chronicles of early Melbourne 1835 to 1852*. Melbourne, 1888.

Le Souef, W. Dudley, 'Acclimatisation in Victoria', *Australian Association for the Advancement of Science, Vol.* II, Melbourne, 1890.

Thomson, Geo M, *On some aspects of acclimatisation in New Zealand*, Australian Association for the Advancement of Science, Vol.III, Christchurch, 1891.

Wilson, Edward, *Acclimatisation*, a paper read before the Royal Colonial Institute. London, 1875.

Wilson, Edward, 'On the Murray River cod, with particulars of experiments instituted for introducing this fish into the River Yarra-Yarra', *Transactions of the Philosophical Institute of Victoria*, (Melbourne,1857), p.25.

—'On the introduction of the British Song Bird', *Transactions of the Philosophical Institute of Victoria*, (Melbourne,1857), pp.77-88.

LATER WORKS

Books

Adelaide Zoo. Revised edition, Adelaide, 1984.

Allen, David Elliston, *The naturalist in Britain : a social history.* London, 1976.
Altick, Richard D., *The shows of London.* Cambridge, Mass., 1978.
Australian Dictionary of National Biography. Melbourne, 1966-
Barber, Lynn, *The heyday of natural history. 1820-1870.* London, 1980.
Berger, John, *About looking.* New York, 1980.
Berman, Morris, *Social change and scientific organization : the Royal Institution, 1799-1844.* Ithaca, 1978.
Blunt, Wilfred, *Ark in the park, the Zoo in the nineteenth century.* London, 1976.
Boussiac, Paul, *Circus and culture: a semiotic approach.* New York, 1985.
Brockway, Lucile H., *Science and colonial expansion: the role of the British Royal Botanic Gardens.* New York, 1979.
Cherfas, Jeremy, *Zoo 2000, a look beyond the bars.* London, 1984.
Clark, Ronald, *The survival of Charles Darwin.* London, 1984.
Darwin, Charles, *The origin of the species.* New York, 1958.
de Serville, Paul, *Port Phillip gentlemen and good society in Melbourne before the gold rushes.* Melbourne, 1980.
Davison, Graeme, *The rise and fall of marvellous Melbourne.* Melbourne 1978.
Eco, Umberto, *Travels in hyper reality.* San Diego, 1986.
Grant, James & Serle, Geoffrey, *The Melbourne Scene, 1803-1956.* Marrickville, NSW, 1978.
Hancocks, David, *Animals and architecture.* New York, 1971.
Hediger, H., *Wild animals in captivity.* New York, 1964.
Henty, Carol, *For the people's pleasure: Australian botanic gardens.* Richmond, 1988.
Home, R W (ed.), *Australian science in the making.* Cambridge, 1988.
Jenkins, C F H., *The Noah's ark syndrome: one hundred years of acclimatization and zoo development in Australia.* Perth, 1977.
Kynaston, Edward, *A man on edge : a life of Baron Sir Ferdinand von Mueller.* London, 1981.
Midgley, Mary, *Animals and why they matter.* Harmondsworth,1983.
Mitchell, P.C., *Centenary history of the Zoological Society of London.* London, 1929.
Moyal, Ann Mozley (ed.), *Scientists in nineteenth century Australia : a documentary history.* Melbourne, 1976.
Mullan, Bob and Marvin, Garry, *Zoo culture.* London, 1987.
Mumford, Lewis, *The city in history: its origins, its transformations, and its prospects.* London, 1861.
Pescott, R T M., *The Royal Botanic Gardens Melbourne: a history from 1845-1870.* Melbourne, 1982.
Prince, J. H., *The first one hundred years of the Royal Zoological Society of N.S.W. 1879 to 1979.* Sydney, 1979?

Regan, Tom, and Singer, Peter (eds.), *Animal rights and human obligations*. Englewood Cliffs,N.J., 1976.
Rix, C E., *Royal Zoological Society of South Australia, 1878-1978*. Adelaide, 1978.
Rolls, Eric, *They all ran wild: the animals and plants that plague Australia*. Sydney, 1984.
Reingold, Nathan, and Rothenberg, Marc (eds.), *Scientific colonialism: a cross-cultural comparison*. Washington D.C., 1987.
St Leon, Mark, *The Circus in Australia*. Wahroonga, N.S.W., 1981.
Scherrin, H., *The Zoological Society of London*. London, 1905.
Serle, Geoffrey, *The Golden age : a history of the colony of Victoria, 1851-1861*. Melbourne, 1977.
— *The rush to be rich*. Melbourne, 1971.
Singer, Peter (ed.), *In defence of animals*. Oxford, 1985.
Thomas, Keith, *Man and the natural world*. Harmondsworth, 1984.
Turner, James, *Reckoning with the beast*. Baltimore, US, 1980.
Walker, Jean, *Origins of the Tasmanian trout*. Hobart, 1988.
Zuckerman,S. (ed.), *The Zoological Society of London 1826-1976 and beyond : the proceedings of a symposium*. London, 1976.

Articles

Byrne, Gerald, 'The experimental farm in the Royal Park'. *Victorian Historical Magazine*, XV (4), 1935.
Desmond, Adrian, 'The making of institutional zoology in London 1822-1836', *History of science*, XXIII, 1985.
Gill, Edmund D., 'Contribution to science by early geologists of FNCV', *Field Naturalist*, 97, 1980.
Gillbank, Linden, 'The Acclimatisation Society of Victoria', *Victorian Historical Journal*, 51 (4), 1980.
Gillbank, Linden, 'The origins of the Acclimatisation Society of Victoria : practical science in the wake of the goldrush' *Historical Records of Australian science*, VI (3), 1986.
Hoare, M.E., 'Learned societies in Australia: the foundation years in Victoria, 1850-1860', *Records of the Australian Academy of Science*, 1 (2), 1967.
Home, R.W., 'Australian science and its public', *Australian Cultural History*, no. 7,1988.
Kohlstedt, Sally Gregory, 'Australian Museums of natural history : public priorities and scientific initiatives in the 19th century', *Historical records of Australian science*, V, (4), 1983.
le Souef, J Cecil, 'Acclimatisation in Victoria', *Victorian Historical Magazine*, 36, 1965.
Osborne, Michael A., 'A collaborative dimension of the European empires : Australian and French acclimatization societies and intercolonial scientific cooperation', 1988?

unpublished paper [acquired from the Dept of the History and Philosophy of Science, University of Melbourne]

Powell, J.M., 'A baron under siege: von Mueller and the press in the 1870s' *Victorian Historical Journal*, 50, (1), 1979.

— 'Melbourne and Kew : Botanic Gardens controversies in the 1870s' *Landscape Australia*, no. 1, 1979.

— 'The botanical Baron and the colonial scribes : Botanic Gardens controversies in the 1870s', *Landscape Australia*, nos 2-3, 1979.

Sanderson, W.A., 'Royal Park', *Victorian Historical Magazine*, XIV (3), 1932.

Willis, J.H., 'The first century of the Field Naturalists Club of Victoria', *Victorian Naturalist*, 97, (3).

CD-ROM CONTENTS

Chronology of Melbourne Zoo from the minutes, 1861 – 1964

A single PDF file (**Chronology.pdf**), providing a searchable summary of key events recorded in the minutes of the Melbourne Zoo's governing body, variously named *Acclimatisation Society of Victoria, 1861-1872; Zoological and Acclimatisation Society of Victoria, 1873- 1937; Zoological Board of Victoria, 1937- 1964*

Maps and Photographs, 1858 – 1951

A selection of images from the Annual Reports, and various other sources as listed below.
The index to the images is in PDF format (**Maps and Photographs.pdf**); to view images, click on the links provided. You will need a browser capable of opening tiff files.

Date	Description
1858	Zoological Society of Victoria, *Rules and Regulations ...*
1858	'Plan of the Zoological Gardens, Richmond Paddock', *Victorian Agricultural and Horticultural Gazette*, April 1858.
1864	Detail, showing location of aviary and 'enclosure for animals', from 'Plan of the Government House Reserve, Botanic Garden and its Domain', in *Annual Report of the Government Botanist and Director of the Botanic Garden*, Victorian Parliamentary Papers 1864-65.
1875	'Guide to the Zoological and Acclimatisation Society's Gardens'
1877	'The New Lion House, Royal-Park', *Australasian Sketcher* (Melbourne), 20 January 1877
1880	'Acclimatization Society's Grounds', in Walch, *Victoria in 1880* (Melbourne: George Robertson, [1880])
1882	'The Indian Animals at the Royal-Park', *Australasian Sketcher* (Melbourne), 6 May 1882
1882	'The Native Encampment, Zoological-Gardens, Royal-Park', *Australasian Sketcher* (Melbourne), 26 August 1882
1884	'Plan of the Gardens of the Zoological and Acclimatisation Society of Victoria, Royal Park, Melbourne'
1886	'The Zoological Gardens' [wood cuts, showing elephant walk], in *Picturesque Atlas of Australasia* (Sydney: Picturesque Atlas Publishing Co., 1886-89])
1888	'In the Zoological Gardens, Royal Park', in Sutherland, *Victoria and its Metropolis: Past and Present* (Melbourne: McCarron, Bird & Co., 1888)
1893	*Guide to the Zoological and Acclimatisation Society's Gardens*

1894	'Lion and Lioness'.
1894	'Tree Kangaroo – "Dendrolagus bennettianus", North Queensland'.
1895	'The Zebra'
1896	'1. Red and White (Albino) Kangaroos (Macropus rufus); 2. Tree Kangaroos (Dendrolagus bennettianus); 3. Indian Elephant; 4. View of Owl House'.
1899	'View in the Gardens'
1901	'Carnivora Exercising Yard'
1901	'Main Walk at Entrance'
1901	'Silver Gull Enclosure'
1908	'Song Bird Aviary, Melbourne Zoological Gardens'
1908	'Flight Aviary, Melbourne Zoological Gardens'
1908	'Ourang-Outang [i.e. Orang Utan] ... Eight Years in the Gardens'
1909	'View in Flight Aviary, Melbourne Zoo'
1911	'Adjutant'
1911	'Mammal Houses'
1911	'Mammal Houses' [another view]
1911	'Shelter Shed'
1913	'Flight Aviary'
1913	'North American Bison'
1913	'Dromedary and Young'
1913	'Elephant Fountain'
1913	'Flamingo Pond'
1913	'View in the Gardens'
1913	'New Giraffe Enclosure'
1913	'Heron Enclosure'
1913	'Hippopotamus'
1913	'Orang-utan'
1913	'Polar Bears'
1913	'Song-bird Aviary'
1913	'Thars and Fat-tailed Sheep'
1913	'Bengal Tiger'
1913	'Waterfowl Pond'
1914	'Angora Goat's Enclosure'
1914	'Livingstone Eland'
1914	'Hippopotamus Enclosure'

1914	'Zebu Cattle Enclosure'
1916	'Young Tibetan Bears and Mother'
1916	'Hippopotamus'
1917	'Fat-tailed Sheep and House'
1918	'Giraffe (can reach up to 15ft)'
1918	'Hippopotamus, Mother and Young, (1¼ Years)'.
1919	'Giraffe (15 ft in height)'
1919	'Ourang-Outang (Mollie)'
1919	'Thars'
1920	'Hippopotamus with Young'
1920	'African Lion'
1922	*Official Guide to the Gardens of the Royal Zoological and Acclimatisation Society of Victoria*
1925	'New Bear Enclosures'
1925	'New Refreshment Kiosk'
1925	'Electric Tram and Railway Entrance'
1926	'Bear Enclosures'
1926	'Monkey Pavilion'
1926	'"Jimmy" the Orang-Utang'
1927	'Large Carnivora Yards, East View'
1927	'Large Carnivora Yards, South View'
1927	'"Jimmy" [orang utan] with Kittens'
1927	'Small Carnivora Yards, East View'
1928	'Young Giraffes'
1928	'Section of Large Carnivora Yards'
1928	'House for Small Mammals'
1929	'African Crowned Crane'
1929	'Tenassirim Muntjac, or Barking Deer'
1930	'European Flamingo and Demoiselle Crane'
1930	'Entellus Monkey Pavillion'
1930	'Rough Billed Pelican'
1934	'General View of the New Australian Section. "Platypussary" in centre. Koala reserve right centre. Dingo yards left centre. Echidna run left foreground. Wombat yards right foreground. Bird aviaries background'
1937	Photograph of Monkey Island, front cover of publicity leaflet
1951	'Zoological Gardens as at 8/10/51' [map]